ACTIVITY-BASED TUTORIALS

Volume 1

Introductory Physics

THE PHYSICS SUITE

Michael C. Wittmann
University of Maine

Richard N. Steinberg
The City College of New York

Edward F. Redish
University of Maryland

and the
University of Maryland
Physics Education Research Group

JOHN WILEY & SONS, INC.

Cover image: James Fraher/Image Bank/Getty Image

To order books or for customer service, please call 1-800-CALL-WILEY (225-5945).

ISBN 0-471-48776-7

Printed in the United States of America

10 9 8 7 6 5 4 3 2 1

Printed and bound by Malloy Lithographing, Inc.

Table of Contents

Activity-Based Tutorials

I. Introduction

For this and all tutorials, remember to work in groups! Everyone in the group should discuss and agree on an answer before moving on. If someone has a question, the whole group should work together to address it.

In this tutorial, you will use a motion sensor attached to a computer. The motion sensor is able to measure distance to the nearest object in front of it. It can also use this data to calculate and display the object's velocity and/or acceleration as a function of time.

A. Set up the motion sensor and the space around your work area so that you have room to walk back and forth for up to 3 meters and the motion sensor can see you throughout your path without interference from other objects.

B. Click the *Start* button to measure distances using the motion sensor. Stand about 50 cm from the ranger and start collecting data. Watch the graph on the computer. Move slowly back and forth to see how the computer displays your position.

Everyone in your group should take turns using the computer for the following activities.

> **Using the motion sensor**
>
> a. The ranger will only measure distances greater than about 10 cm (4 inches). If you get too close, the ranger will give you the wrong distance.
>
> b. The ranger must be pointed directly at the object it is measuring. It often helps to hold a notebook in front of you as you walk, so that the ranger points at a flat surface.
>
> c. If the object you are measuring gets close to another object (such as a table or one of your partners) it may measure the distance to that object instead of the one you want.

II. Making a Distance Graph

In this section, you will make a number of distance vs. time graphs. In order for this section to be effective, you must rotate who runs the computer and who does the walking through your working group. Change after ***every*** trial.

A. Some Simple Distance Graphs

In the space below you are asked to make predictions **before** carrying out an experiment. Draw each prediction as a dotted line on each graph. Do each experiment. Draw the resulting graph as a solid line. (Do not erase your prediction!) Be sure to label units on the graph axes in your sketch.

1. Predict what the distance/time graph will look like if you start at the 0.5 m from the ranger and walk away from the motion sensor (origin) *slowly and steadily.* Carry out the experiment.

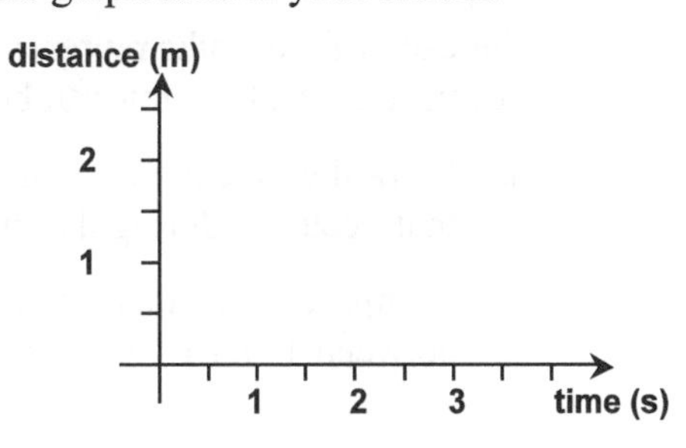

2. Predict what the distance/time graph will look like if you walk away from the motion sensor (origin) *medium fast and steadily*. Carry out the experiment. (Remember to rotate who controls the computer and who does the walking.)

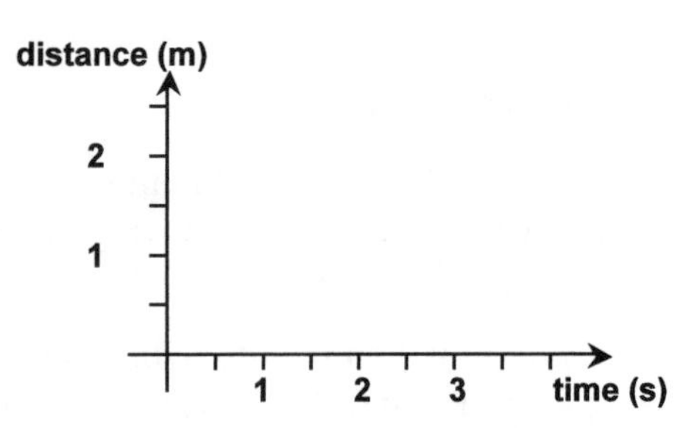

3. Predict what the distance/time graph will look like if you walk towards the motion sensor (origin) *slowly and steadily*. Carry out the experiment. (Remember to rotate who controls the computer and who does the walking.)

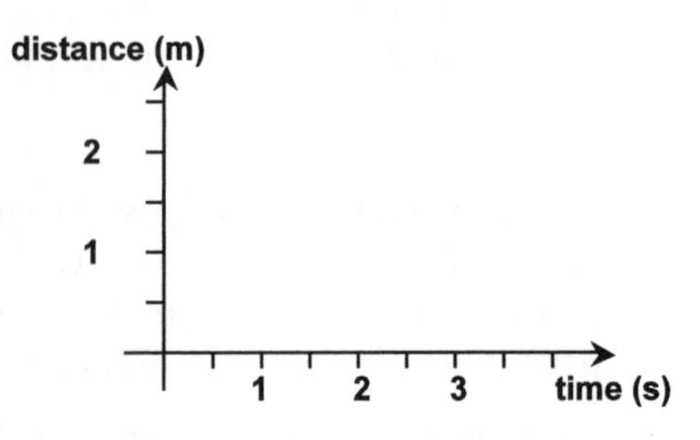

4. Predict what the distance/time graph will look like if you walk towards the motion sensor (origin) *medium fast and steadily*. Carry out the experiment. (Remember to rotate.)

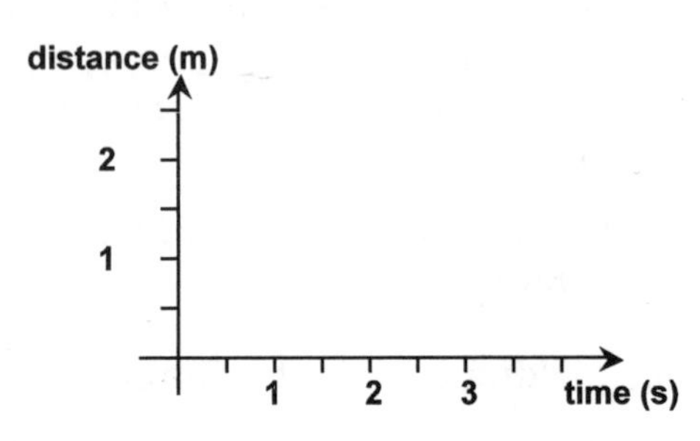

5. Describe the difference between the graphs you made by walking away slowly and the graphs you made by walking more quickly.

6. Describe the difference between the graphs you made by walking toward the motion sensor and the graphs you made by walking away from the motion sensor.

7. For each of the walks represented in graphs 1–4 above, estimate the displacement that occurred in the time interval between 1 and 1.5 s after the data collection started.

   a. Mark the time interval on each graph and the displacement that occurred during that time interval.

   b. Compare the displacements that occurred in the interval between 1 and 1.5 s for the four cases.

   c. In each case, would the displacements have been different if you considered the same time interval but occurring later in the motion? Explain.

B. A More Complex Distance Graph

1. For this question, work individually. Consider the following motion: A person walks away from the motion sensor slowly and steadily for 4 seconds, stops for 4 seconds, then walks toward the motion sensor quickly. Use a dotted line to draw your prediction of the distance vs. time graph of this motion on the left graph below.

2. Now compare your predictions with the other members of your group. If you have differences, discuss them and see if you can all agree. Draw your group prediction on the left graph using a solid line. (Do not erase your original prediction!)

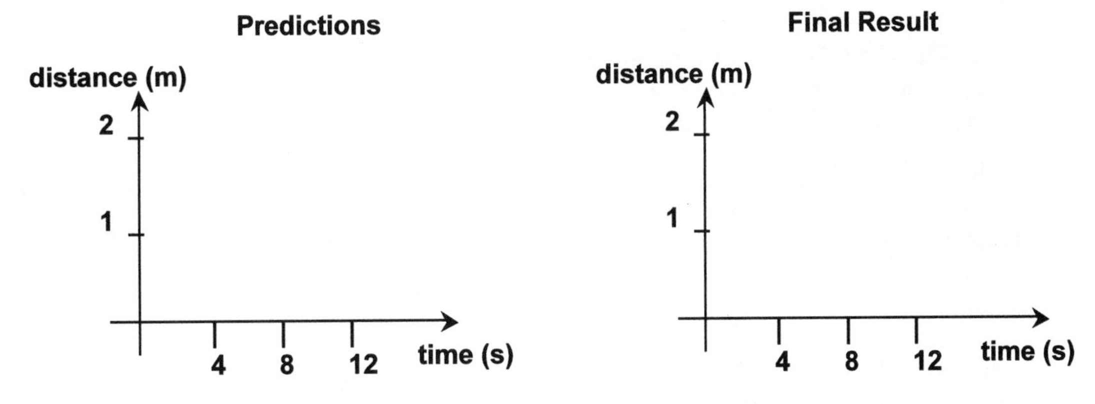

On your computer, change the scale on the time axis to 12 sec. Now do the experiment. When you are satisfied with your graph, draw your group's final result on the right graph.

3. How did you determine if you were "satisfied" with your graph? Explain.

4. Is your prediction the same as the final result? If not, describe how you would move to make a graph that looks like your prediction.

5. How would you infer from the graph on the right what your velocity was at different times during the motion? Explain.

6. What characteristics of the graph can you use to infer:

   a. the velocity at each instant of time?

   b. the average velocity for the *entire* motion?

III. Making Velocity Graphs

In this section, you will have the computer plot your velocity rather than your position. Set your computer to display a velocity graph and not a position graph. Scale your velocity graph so that the velocity scale ranges from –1 m/s to +1 m/s and the time axis ranges from 0 s to 4 s.

A. Some Simple Velocity Graphs

Predict what the graph will look like before doing the experiment. Draw your prediction as a dotted line. After you have done the experiment, draw the result as a solid line. (Do not erase your prediction!)

1. Predict what the velocity/time graph will look like if you walk away from the motion sensor (origin) *slowly and steadily*. Carry out the experiment.

   At what velocity did you walk? Use your answers to section II to arrive at an answer. (If you have not already done so, label the values of your graph.)

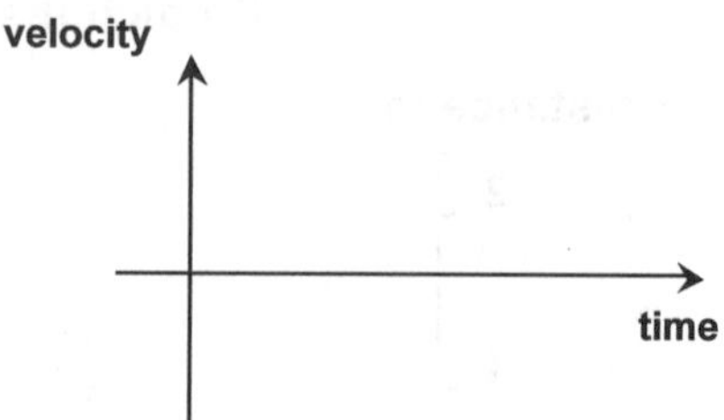

2. Predict what the velocity/time graph will look like if you walk away from the motion sensor (origin) *medium fast and steadily*. Carry out the experiment.

   At what velocity did you walk? Explain.

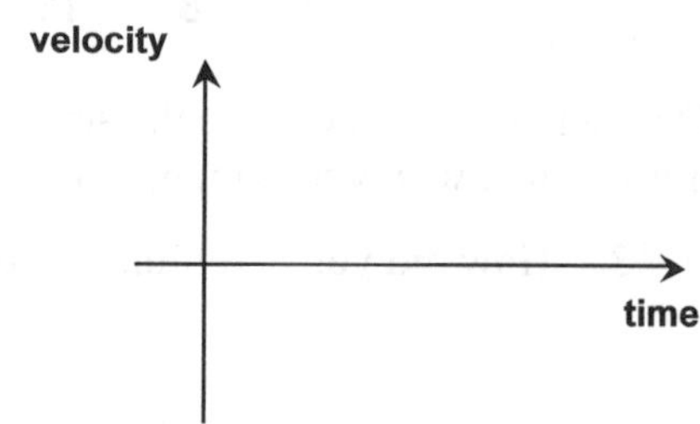

3. Predict what the velocity/time graph will look like if you walk towards the motion sensor (origin) *slowly and steadily*. Carry out the experiment.

   At what velocity did you walk? Explain.

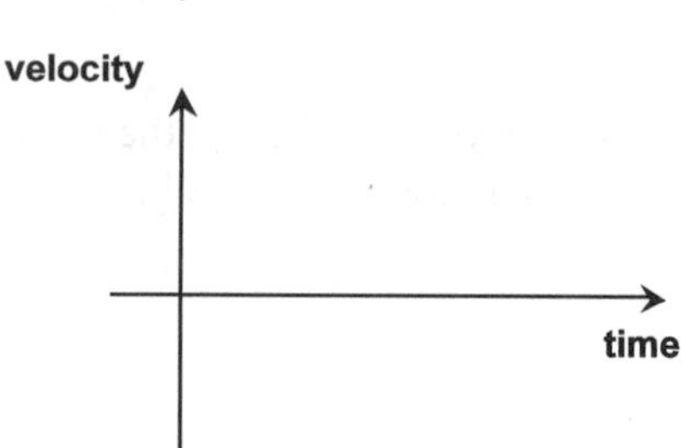

4. Predict what the velocity/time graph will look like if you walk towards the motion sensor (origin) *medium fast and steadily*. Carry out the experiment.

   At what velocity did you walk? Explain.

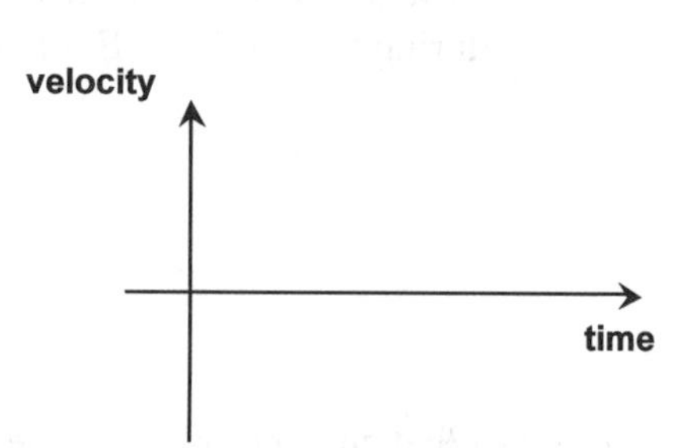

5. What is the difference between the graphs you made by walking away slowly and the graph you made by walking more quickly?

6. Describe the difference between the graphs you made by walking toward the motion sensor and the graphs you made by walking away from the motion sensor.

7. Would the graphs be different if you moved in the same way but in each case started closer to the motion sensor? If so, how? Explain.

B. More Complex Velocity Graphs

1. For this question, work individually. Consider the following motion: A person walks away from the motion sensor slowly and steadily for 10 seconds, stops for 4 seconds, then walks toward the motion sensor steadily about twice as fast as before. Use a dotted line to draw your prediction of the distance vs. time graph of this motion on the left graph below.
2. Now compare your predictions with the other members of your group. If you have differences, discuss them and see if you can all agree. Draw your group prediction on the left graph using a solid line. (Do not erase your original prediction!)

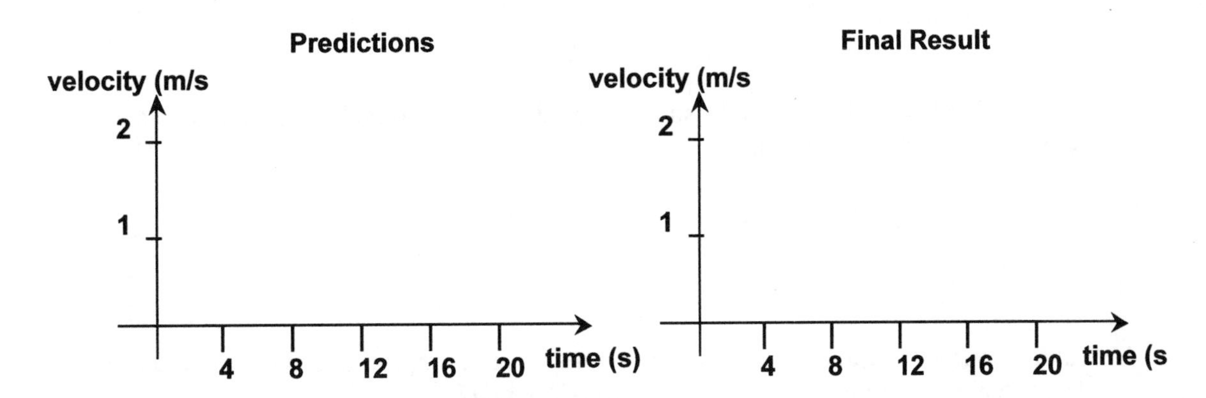

3. Adjust the time scale to 20 seconds. Now do the experiment. Repeat your motion until you think it matches the description. Draw the best case on the right graph above.
4. Is your prediction the same as the final result? If not, describe how you would move to make a graph that looks like your prediction.

C. Matching a Velocity Graph

Consider that you are given a velocity graph and must match your motion to the motion indicated on the graph below.

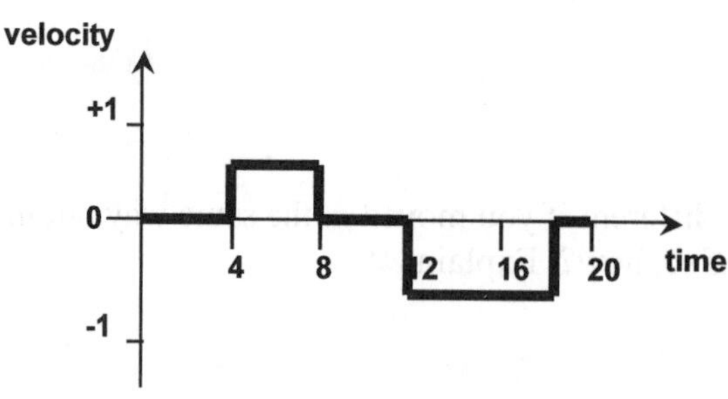

1. Describe how you think you have to move in order to match this graph. (Make sure your whole group agrees on the answer.)

2. Start collecting data and move so as to match the graph shown. You may try a number of times. Get the times and velocities right. *Each person* in the group should take a turn. Draw your group's best effort on the graph above.

3. Is it possible for an object to move so that it produces an absolutely vertical line in a velocity/time graph? Explain.

4. Taking your answer to question 3 into account, make the appropriate approximations to describe how you moved to match each part of the graph.

5. Did you have any difficulties with your location in any of your trials? For example, did you find yourself coming too close to the ranger at the end?

6. Could you have decided on a place to start walking that would have permitted you to walk the entire motion within the range of the motion sensor without trial and error before you began to walk? How?

I. Motion of a Fan Cart Ignoring Friction

In the following experiments, you will use the motion sensor and a low friction cart. The positive direction is away from the rangers. Slide the carts **gently** on the track to get a feeling for how they move. **Do not drop them or run them off the edge of the table!**

In part I of this tutorial we will ignore friction.

A. Example 1: A fan cart moving in one direction

Turn on the fan but put your hand in front of the cart so that it does not move.

1. Draw a free body diagram for the cart/fan system. Label all the forces acting on the cart/fan by identifying:

   a. the type of force (if possible),

   b. the object on which the force is exerted, and

   c. the object exerting the force.

2. How, if at all, would your free body diagram change if you were to remove your hand from the front of the cart?

3. Suppose that the cart moves away from the motion sensor, starting from rest, while the fan supplies a constant force. Predict what the velocity vs. time and the acceleration vs. time graphs would look like. Explain how you arrived at your answer.

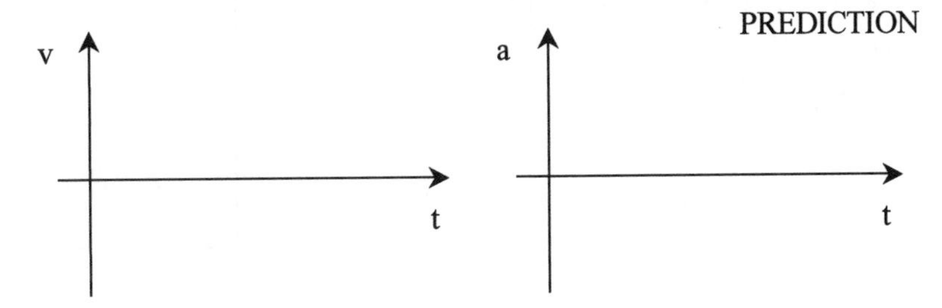

Place the motion sensor so that it measures the velocity of the cart on the table.

4. Let the cart be at rest approximately 0.2 m from the motion sensor. Turn on the fan. Take data using the motion sensor starting at the moment the cart begins to move. Sketch the velocity graph on the axes at right.

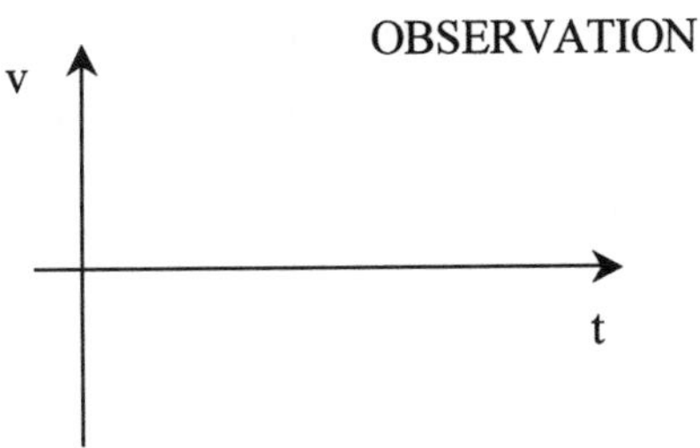

5. Compare your observed graph in question 4 with your prediction in question 3. Resolve any discrepancies.

6. Use the velocity graph to describe the acceleration of the cart. Explain how you arrived at your answer.

7. By clicking your mouse in the window, you can find the value of the velocity at different times. Use these values to find the *average* acceleration of the cart during its motion. Explain how you arrived at your answer.

B. Example 2: A fan cart given an initial push in the same direction as the force of the fan on the cart/fan system

Suppose that you give the cart a small initial push in the same direction as the force exerted on cart/fan system by the fan. In such a situation, the cart/fan system is already moving away from the motion sensor at time t = 0 sec.

1. Draw a free body diagram for the cart/fan system after it is no longer touching your hand. Label all forces on the cart/fan system by identifying:

   a. the type of force (if possible),

   b. the object on which the force is exerted, and

   c. the object exerting the force.

2. Predict what the velocity vs. time and the acceleration vs. time graphs would look like. Explain.

PREDICTION

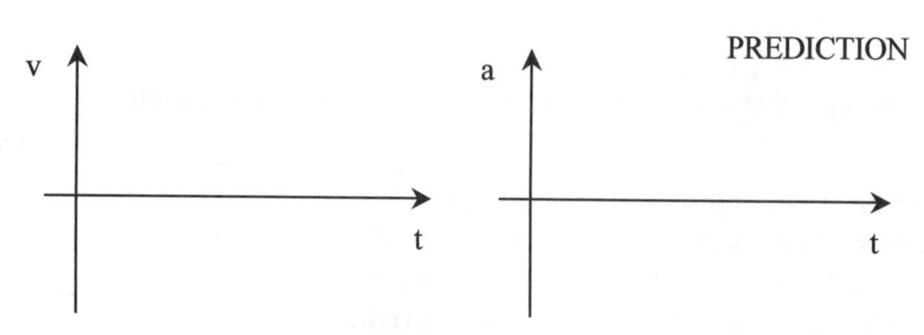

3. How, if at all, do you expect the graphs to be different from the graphs you obtained in part A, Example 1? Explain the reasoning for any of the differences that you expect.

4. Take data using the motion sensor for the situation above. (Give the fan cart a gentle push *before* you start the motion detector.) Sketch the velocity graph in the space to the right.

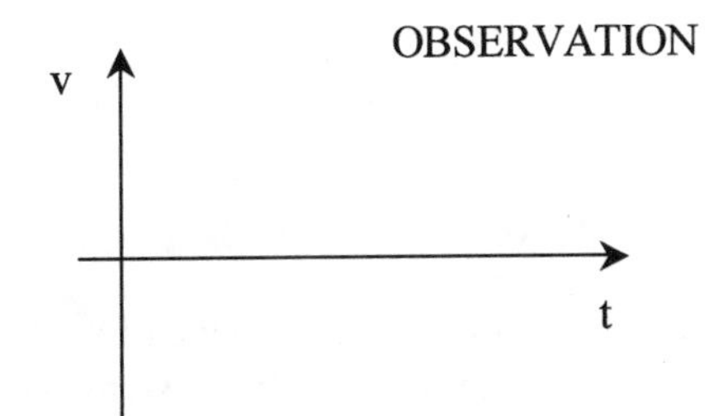

5. Compare your graph with your prediction. Resolve any discrepancies.

6. Compare the accelerations in the graph above to the acceleration in the graph in part A, Example 1. Discuss similarities and differences.

C. Example 3: A fan cart given an initial push in the direction opposite that exerted by the fan on the cart/fan system.

Suppose that you push the cart so that the initial velocity is in the opposite direction of the force the fan exerts on the cart/fan system and towards the motion sensor. Assume that friction is negligible.

1. Draw two free body diagrams below after the cart/fan system is no longer touching your hand. On the left, draw one for the cart/fan system on its way toward the motion sensor. On the right, draw one for the cart/fan system after it turns around.

2. Predict the shape of the velocity and acceleration graphs for this situation. Sketch your predictions on the graphs below.

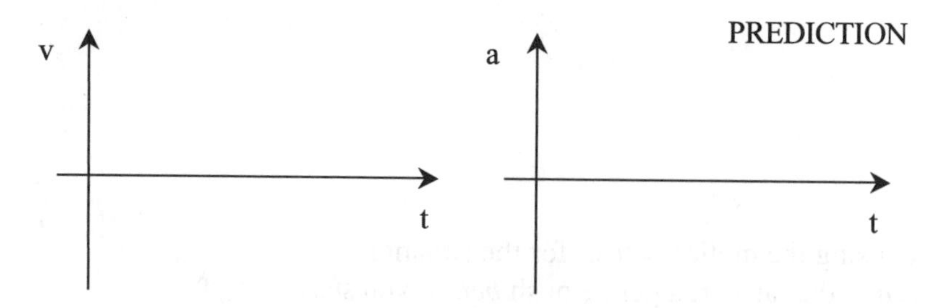

II. Motion of a Fan Cart with Friction

Suppose that you push the cart so that the initial velocity is in the opposite direction of the force the fan exerts on the cart/fan system and towards the motion sensor. However, now do *not* assume that the friction is negligible.

1. Draw two free body diagrams below for the cart/fan system after it is no longer in contact with the hand. On the left, draw one for the cart/fan system on its way toward the motion sensor. On the right, draw one for the cart/fan system after it turns around.

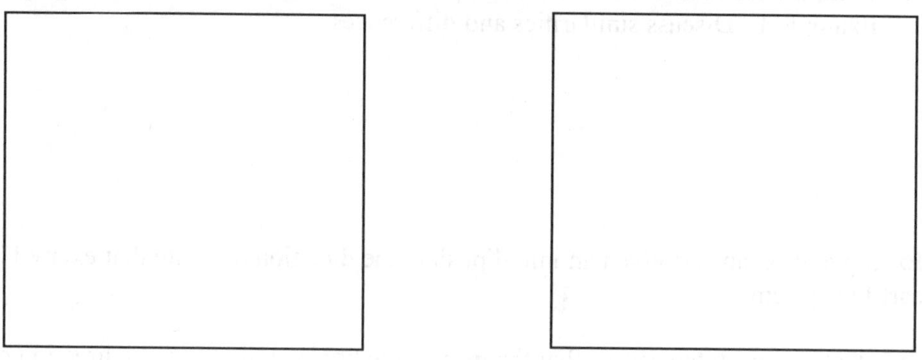

2. Predict the shape of the velocity and acceleration graphs for this situation. Sketch your predictions on the graphs below.

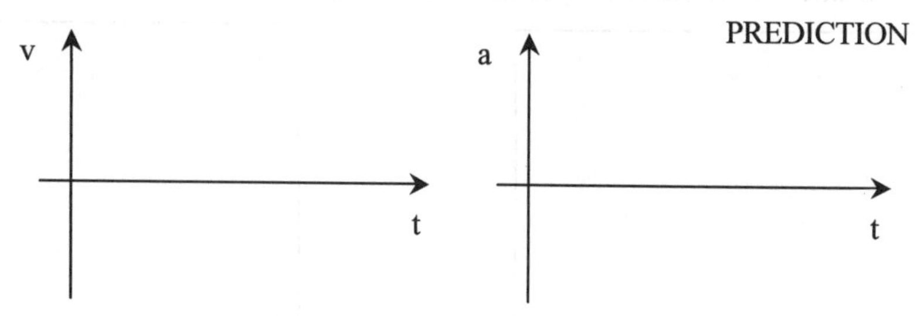

3. Compare these velocity and acceleration graphs to your graphs when there was no friction. Explain any differences between the graphs.

4. Take data using the motion sensor for this situation. You may need to adjust the time axis to display up to 10 seconds. Sketch the velocity graph in the space to the right.

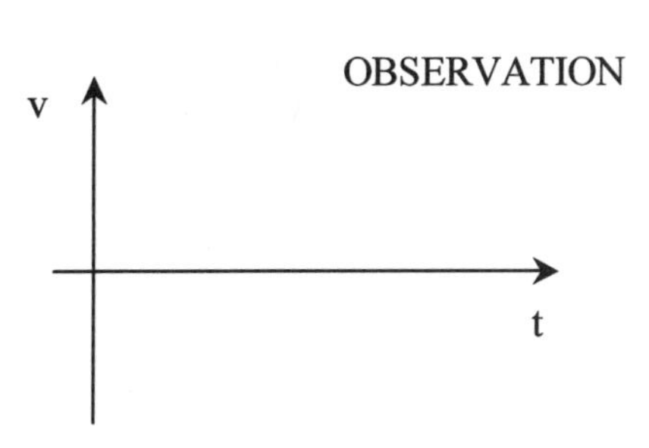

5. Is there evidence that there is a frictional force? Explain.

6. Estimate the value for the frictional force. Explain your reasoning. (A balance scale is provided in the room.)

I. Introduction to the Force Probes

For this tutorial, you will use force probes attached to carts which are able to collide with each other.

A. When using the force probes keep the following in mind:

1. Too much force will damage the force probes, so be careful not to exceed 50 N.
   DO NOT PRESS OR PULL HARD ON THE PROBES.

2. Occasionally, the "zero force" measure of the probe will drift. Sometimes, the force probes must be zeroed again. To do so, click on the *Zero* button on the probe. Make sure there are no pushes or pulls on the force probe at this time.

B. Get a feel for the magnitude of the forces. Have each member of your group push and pull gently on the tip of the force probe and watch the result on the computer graph. Try it first with one probe and then with both.

1. Describe your results.

2. Do the probe forces have the same sign for pushing and pulling?

The sensors for the force probes have been mounted on low friction carts.
Slide the carts **gently** on the table to get a feeling for how they move.

II. Pushing While Not Moving

A. Consider what will happen when we push cart 1 into cart 2 (don't do it yet!).

1. Are both carts exerting forces on each other?

2. If you stated that the carts were exerting forces on each other (meaning that each cart had a force exerted on it), compare the relative magnitude of the forces. If you stated that there there is only one force being exerted, state so explicitly.

B. Use the carts and computer to evaluate your answer to question 1. Click "Start" on the computer. Push the carts together with gradually increasing force and watch the result on the computer screen. Does the result match your predictions in question 1? Account for any discrepancies in the space below.

III. Pushing While Moving

A. Consider a small car pushing a large truck and speeding up. Draw free body diagrams for both the small car and the truck below.

Free body diagrams

car	truck

B. Use your diagrams to make predictions in the following questions.

1. Do both the truck and the car exert forces on one another? Explain how you know.

2. What is the relative magnitude of the forces they exert on each other? Explain your reasoning.

3. Is the net force on the car the same as the net force on the truck? Explain your reasoning.

4. Discuss your answers with your group and come to an agreement on the free body diagrams and what you think is happening. Write down the group's prediction (including the free body diagrams) in the space below.

C. We can model this situation with the low friction carts. Put two bar weights into one of the carts and call it cart 2. Call the other cart 1. In our model cart 2 will represent the truck. Cart 1 will be the small car.

1. Predict what the computer would show if you began pushing cart 2 with cart 1 very slowly at first and then faster always keeping contact between the two carts. Sketch your prediction on the axes below.

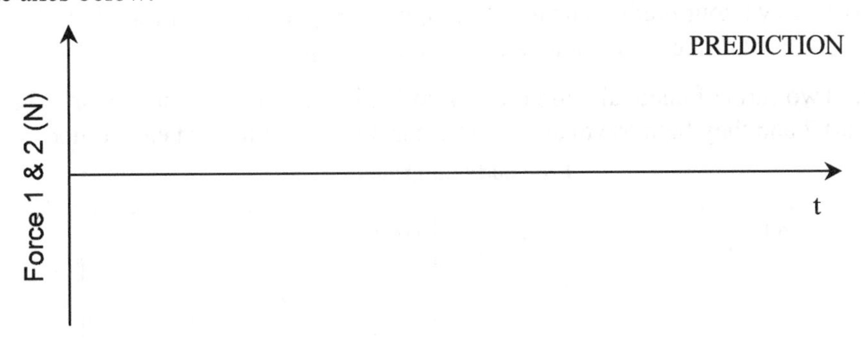

Explain how you arrived at your answer.

2. Using the computer, the carts and the force probes, carry out the experiment. Remember not to exceed 50 N and remember to zero the force probes before taking measurements.

Sketch the graph of the force probe measurements on the axes below.

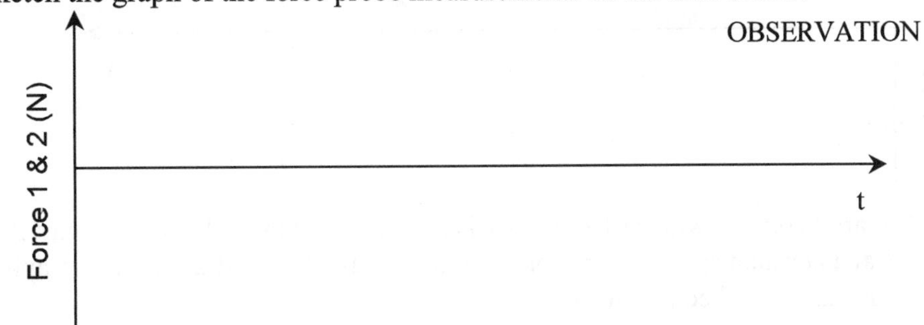

3. Compare your results in question 2 to your prediction in question 1.

   a. Is the magnitude of the force exerted on cart 2 by cart 1 significantly different from the magnitude of the force exerted on cart 1 by cart 2 during any part of the motion?

   b. Explain any differences between your predictions and your observations.

IV. Collisions

Suppose the carts are not continually in contact but collide briefly.

A. Predictions

Consider two situations. For each situation, draw free body diagrams for each cart during the collision, and write down your predictions for the graphs of $\mathbf{F}_{12}$ and $\mathbf{F}_{21}$ as a function of time for the two situations. Use a solid line for $\mathbf{F}_{12}$ and a dashed line for $\mathbf{F}_{21}$.

Situation 1: Two carts of unequal mass moving towards each other collide. Cart 1 has a smaller mass then cart 2 and they both travel at the same constant speed toward each other.

Free body diagrams

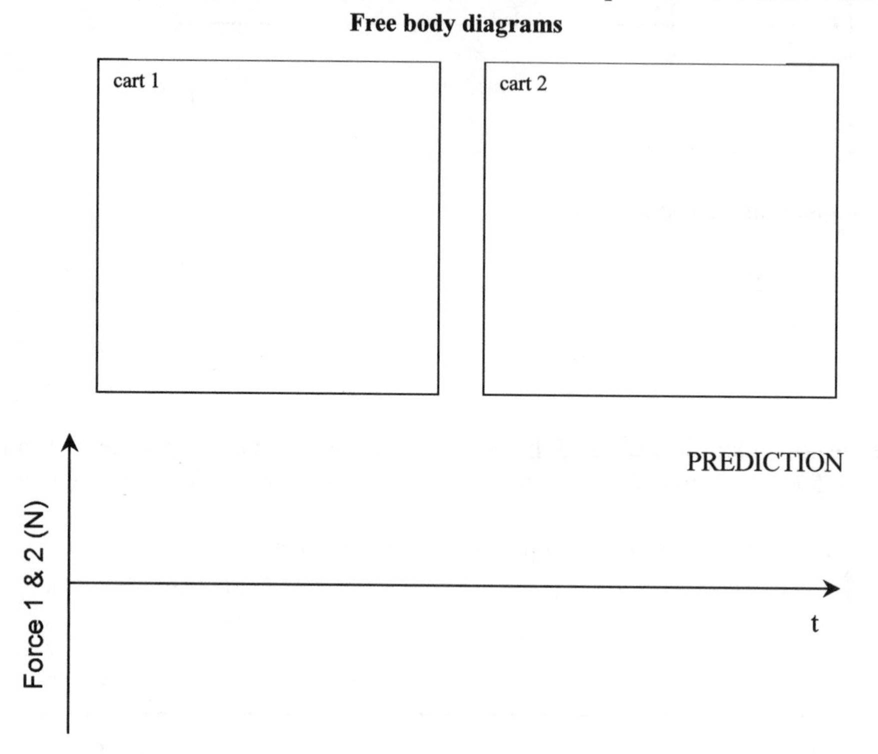

Situation 2. Cart 2 collides with cart 1. Cart 1 is at rest before the collision and cart 2 moves toward cart 1 at a constant speed. Cart 1 has a smaller mass than cart 2. In your graph, use a solid line for $\mathbf{F}_{12}$ and a dashed line for $\mathbf{F}_{21}$.

Free body diagrams

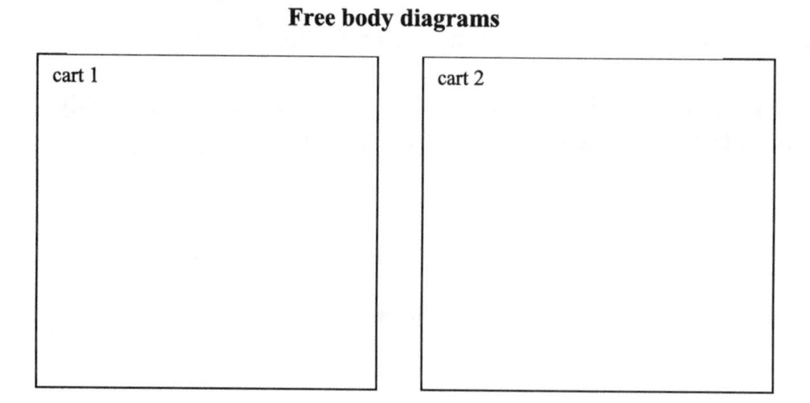

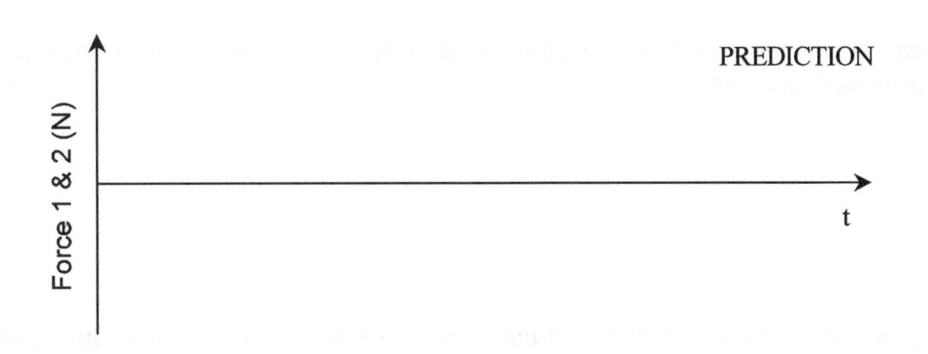

B. Observations

Using settings provided by your instructor, set your computer to wait for triggered data from the force probes.

> Be sure to Zero the force probes before each collision. Make sure no forces are exerted on the probes when you do so. Also remember that the force probe may be damaged by forces larger than 50 N so be careful with the collisions.

1. Use the low friction carts to explore the situations described above and check the results against your groups predictions.
   Sketch the force vs. time graphs below.

Situation 1. $m_2 > m_1$, $|v_1|, |v_2| =$ constant

Situation 2. $m_2 > m_1$, $|v_1| = 0$, $|v_2| =$ constant

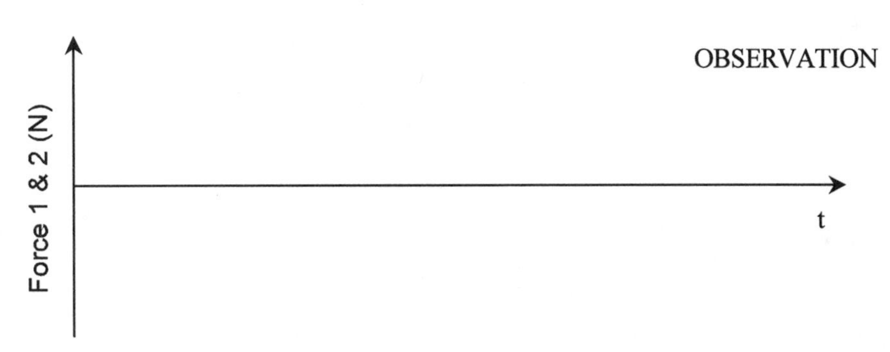

2. How did your observations compare with your predictions? Resolve any discrepancies.

3. Do your observations of the forces make sense in terms of the velocities and accelerations of the two carts? Explain.

4. Did you find a situation where one of the carts exerted a larger force on the other cart?

5. Do you think you can find one? What would this situation be?

6. Explain how your conclusions written above relate to Newton's Laws of motion. Explain and elaborate, giving specific examples and detailed descriptions of the physics.

The Cylinder Cyclone

Direction of spin

Jenny

A. At an amusement park, Jenny decides to take the Cylinder Cyclone ride, shown in the diagram at right. In this ride, a cylinder spins at a constant speed and then suddenly the bottom drops out from underneath everyone's feet. The rotation speed is fast enough that Jenny remains at the same position on the wall, without sliding down.

1. On the top view diagram provided at right, draw arrows representing velocity vectors at the indicated points.

2. Use the vectors to find the direction of Jenny's average acceleration between points A and B. Explain.

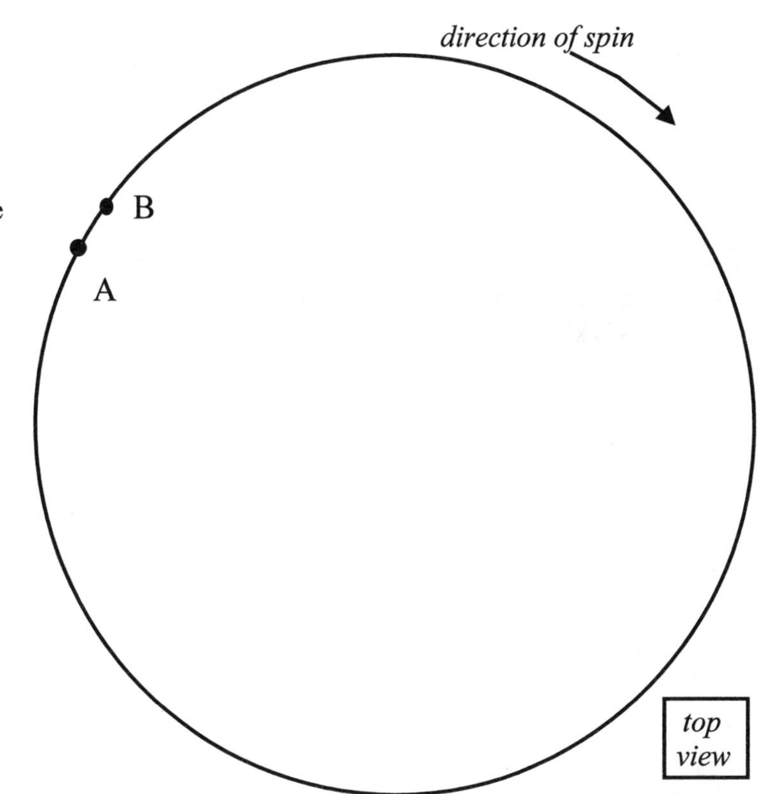

3. Consider the arrows you have used to represent the velocity vectors.

   a. How, if at all, would the acceleration change if Jenny's speed were doubled? Describe both magnitude and direction.

   b. How, if at all, would the acceleration change if the radius of the cylinder were doubled but the speed were held constant? (Consider that the central angle subtended by the arc AB remains the same, not length of AB.)

c. Are your answers to *a* and *b* consistent with the equation $a = v^2/r$? Resolve any discrepancies.

4. Consider that Jenny is located as seen in the first (perspective view) diagram on the previous page, with the wall at her back to the left. Draw a free body diagram for Jenny at this location. Label all forces clearly.

5. What is the work done by *each force* acting on her? Relate your answer to the change in Jenny's kinetic energy. Explain.

6. The knowledgeable ride attendant says that the coefficient of static friction between cotton and the wall of the cylinder is 0.9. If Jenny has a mass of 50 kg, is moving with a speed of 10 m/s, and knows the cylinder is 10 m in diameter, should she worry about slipping when the bottom drops out? Should her boyfriend, who has a mass of 80 kg but is wearing a shirt of the same cotton material? Explain.

B. A small box is given a brief kick and starts with speed v_0 up the incline shown. The incline makes an angle, θ, with the horizontal and has height h. The coefficient of kinetic friction between the box and the incline's surface is μ. The box leaves the ramp at its top, flies through the air, and lands on a table a distance d away from the ramp. Ignore air resistance and the rotation of the box in this problem.

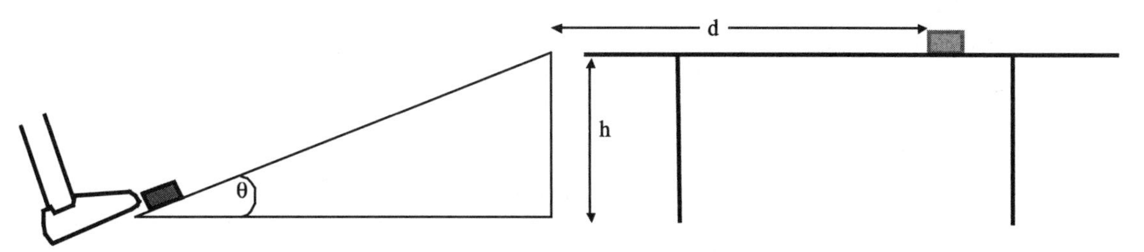

1. Draw free body diagrams for each case below. Label all forces clearly.

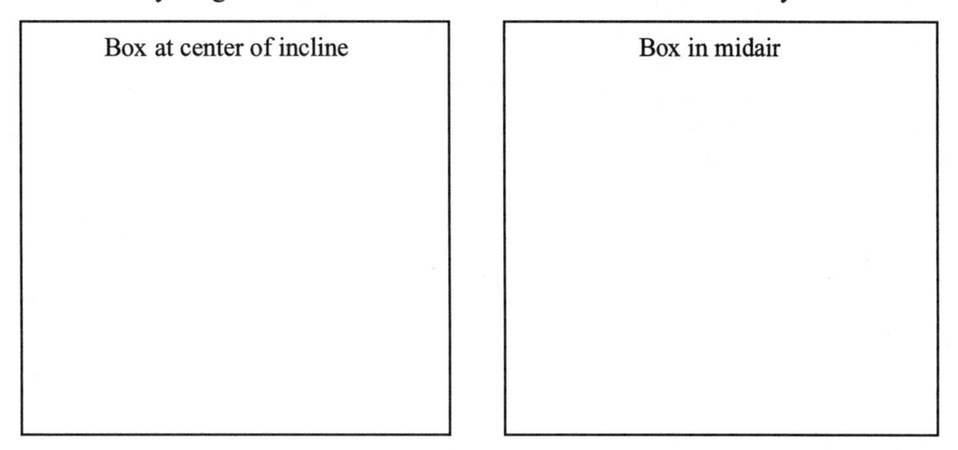

2. Find the work done by *each force* during the entire motion of the box up the incline. Give your answer in terms of the given variables, v_0, θ, h, and μ. Explain how you arrived at your answer.

3. Find the speed of the box at the top of the incline in terms of the given quantities. Explain how you arrived at your answer.

4. Find the distance d in terms of given variables. Show all work.

5. The coefficient of friction, μ, is 0.2, the height of the incline is 1 m, and the angle between the incline and the horizontal is 30°. If the initial velocity of the box were 6 m/s, find the distance d. Show all work.

I. Motion Without Air Resistance

Consider a ball thrown straight upwards. ***For this section, ignore air resistance.***

A. In the box at right, sketch a coordinate system for showing a *picture* of the motion of the ball. Take the starting point to be somewhere near the bottom of the region and make that the origin. Choose the positive y axis to point up. Draw images of the ball:

- after it has left the hand and is part way up
- at the top
- part of the way back down.

B. Suppose the ball is started at time t = 0 with a positive velocity. In the space below, sketch *graphs* that show the height of the ball as a function of time and the velocity of the ball as a function of time. How do you know they look the way you have drawn them?

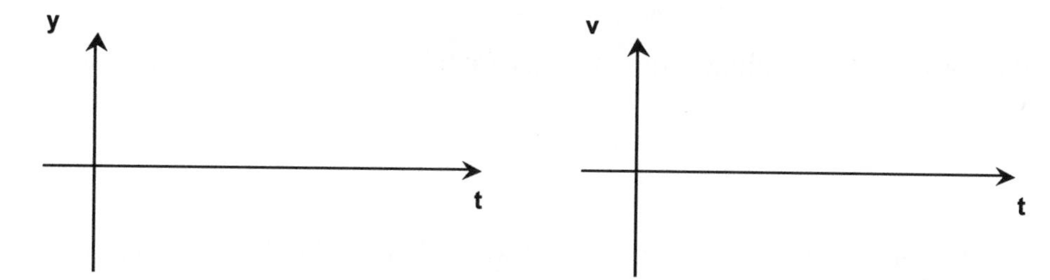

C. On *each* of your images of the ball in question A, draw a free-body diagram indicating all of the forces acting on the ball at that instant of time. Label all forces by identifying: a) the type of force, b) the object on which the force is exerted, and c) the object exerting the force.

D. If you have not done so already, compare the accelerations of the ball (both magnitude and direction) at each of the three locations that you have drawn. Explain your reasoning.

E. Is your answer to question D consistent with the velocity vs. time graph that you sketched? How can you tell? If your answer is not consistent with your graph, resolve the discrepancy.

II. Motion with Air Resistance

A. Thinking about Air Resistance

1. Assume that we have two bodies with different masses and drop them. If we ignore air resistance, what are the forces acting on the bodies while they are falling?

2. Write an equation that gives the response of each body to its net force and show that they will have the same accelerations.

3. In the real world, do all falling bodies have the same acceleration? If not, give a real world example.

We might guess that the reason that bodies in the real world do not all fall in the same way is because something else is acting on them other than the earth. Since we know various kinds of contact forces, we might guess that something is in contact with the objects as they fall—the air—and that exerts a force. We'll have to try to make a guess at what that force is.

4. Would you expect the force that air exerts on an object moving through it to depend on the relative velocity between the object and the air or not? Consider your personal experience with wind.

B. Modeling the Force of Air Resistance

Three models of resistive forces are often used:

- Newton drag (force proportional to v^2)
- viscous drag (force proportional to v)
- friction (force independent of the magnitude of v).

We will consider the behavior of a ball that is feeling a drag force. (This is the most realistic for objects the size of a ball.)

Consider a ball thrown straight upwards.
For this section, suppose that there is a force of air resistance with magnitude $F_{ba} = cv^2$.

1. In the space at the right, sketch a coordinate system for showing a *picture* of the motion of the ball. Draw images of the ball:

- after it has left the hand and is part way up
- at the top
- part of the way back down.

2. In the space below, draw *graphs* that show the height of the ball as a function of time and the velocity of the ball as a function of time. How do you know they look the way you have drawn them?

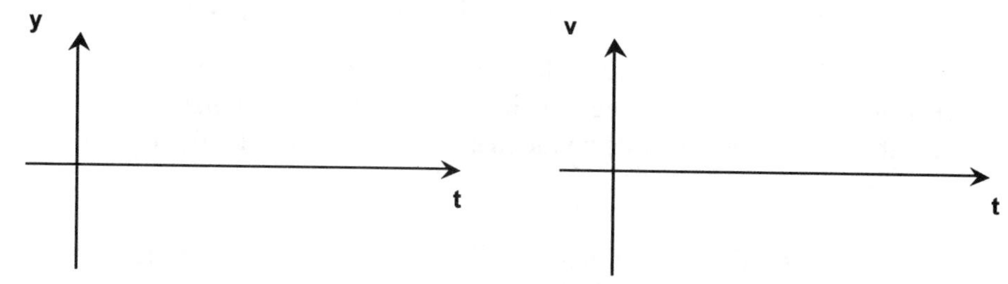

3. On *each* of your images of the ball in question 1, draw a free-body diagram indicating all of the forces acting on the ball at that instant of time. Label all forces by identifying: a) the type of force, b) the object on which the force is exerted, and c) the object exerting the force.

4. If you have not done so already, compare the accelerations of the ball (both magnitude and direction) at each of the three locations that you have drawn. Explain your reasoning.

   a. After reaching its highest point, what happens to the magnitude of the drag force as the ball falls? Explain your reasoning.

   b. What happens to the acceleration as it falls? Explain your reasoning.

c. What happens to the velocity as it falls? At very large times is the velocity changing significantly? Derive an equation for the value at which the velocity flattens out. This is called the *terminal velocity*.

5. Are your answers to question 4 consistent with the velocity vs. time graph that you sketched? How can you tell? If your answer is not consistent with your graph, resolve the discrepancy.

III. Testing Models of Air Resistance

In this part of the tutorial you will need coffee filters and a meter stick. Coffee filters have the property that they are light enough and have a large enough area that they reach terminal velocity almost immediately. This means that if you drop one at a distance of a meter it floats down smoothly at essentially a constant speed. This will allow us to determine whether they feel an air resistance force that is approximately proportional to v^2 (Newton drag) or to v (viscous drag). (If you don't get to this part of the tutorial during the section, you can easily do this part at home.)

We will write equations that express the condition that the filter is falling at terminal velocity and rearrange it to derive a simple expression for the time it takes to fall a certain distance. From this, we will be able to see how the time of fall depends on the mass (number of nested filters) and on the distance.

A. Suppose a filter of mass m is falling at terminal velocity under the influence of an air resistance force $\mathbf{F}_{fa}$. What is the condition that tells you it will be falling at terminal velocity?

B. Suppose a filter of mass m feels a viscous drag force $\mathbf{F}_{fa} = -b\mathbf{v}$. What will be its terminal velocity?

C. If you let a filter fall a distance S, how long will it take to fall? Express your answer in terms of m, g, b, and S. (Assume it is traveling at terminal velocity the whole time. This is a pretty good approximation.)

D. Suppose you could double the filter's mass without changing its air resistance force. How would the time it takes to fall a distance S change?

You effectively double the filter's mass without changing the air resistance it feels by nesting two filters inside each other. Since only one of them is pushing its way through the air, the air resistance force doesn't change.

E. Find a new distance, D, so that the time it takes the double filter to fall D is the same as the time it takes a single filter to fall a distance S?

F. Take S to be 1 meter. Drop two filters one nested inside each other from the height D at the same time as you drop a single filter from the height S. Do they hit at the same time?

G. Repeat the argument assuming quadratic drag, $F_{fa} = -cv^2$. Now what should D be compared to S? If you use the distances given above, do the filters now hit at the same time? Explain your conclusions.

I. Motion Without Air Resistance

Consider a ball thrown straight upwards. ***For this section, ignore air resistance.***

A. In the box at right, sketch a coordinate system for showing a *picture* of the motion of the ball. Take the starting point to be somewhere near the bottom of the region and make that the origin. Choose the positive y axis to point up. Draw images of the ball:

- after it has left the hand and is part way up
- at the top
- part of the way back down.

B. Suppose the ball is started at time t = 0 with a positive velocity. In the space below, sketch *graphs* that show the height of the ball as a function of time and the velocity of the ball as a function of time. How do you know they look the way you have drawn them?

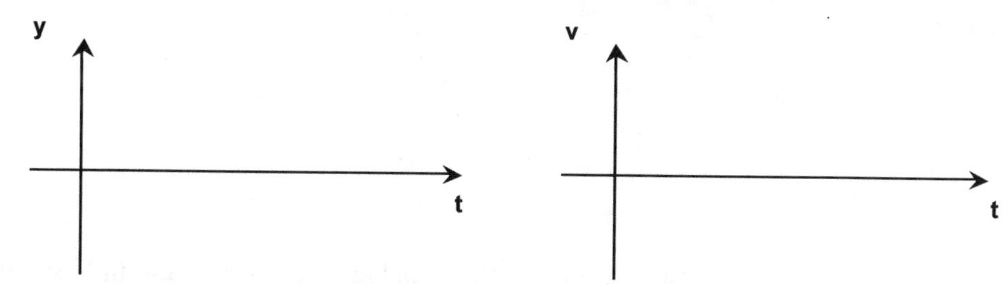

C. On *each* of your images of the ball in question A, draw a free-body diagram indicating all of the forces acting on the ball at that instant of time. Label all forces by identifying: a) the type of force, b) the object on which the force is exerted, and c) the object exerting the force.

D. If you have not done so already, compare the accelerations of the ball (both magnitude and direction) at each of the three locations that you have drawn. Explain your reasoning.

E. Is your answer to question D consistent with the velocity vs. time graph that you sketched? How can you tell? If your answer is not consistent with your graph, resolve the discrepancy.

II. Modeling Motion Without Air Resistance

Open the program `AIRRES1` on your computer. It calculates the motion of a ball in one dimension. To use this program, use the menu bars at the top of the screen. The item highlighted in white is the *selected* item. To change which item is highlighted, use the left mouse button or the left and right arrow keys. To *choose* a selected item, press `<enter>`.

Choose `SET DATA`. You will now see the *data-entry screen.* The numbers the program is currently using are shown in the black boxes. You can change any of them by typing over the numbers that are there. For now, look at the numbers and notice the air resistance is set to zero. Press `<enter>` to accept this data. Notice that there are now two graphs and an animation. Also notice that the menu has changed to the *graph menu.*

A. Choose `PLOT` to watch the graph again. Do the graphs look like what you expected? Explain any inconsistencies.

B. Choose `MEASURE` to measure the total amount of time the ball takes to rise and fall. Point to the place on the position graph that corresponds to the ball returning to its starting point and click with the left mouse button. The x and t values of the point you clicked on will be displayed at the top of the screen. How close to 0 can you get for the x value? Try a number of times. The uncertainty in this measurement gives you an idea of how accurately you can determine a value from this graph.

Write down the total time, Δt, the ball took to rise and fall and your uncertainties in reading x and t which we will call *dt* and *dx*. (Use the notation $\Delta t \pm dt$ where now Δt and dt are both numbers.)

C. Choose `MEASURE` and find the time at which the ball reaches the peak. Within your uncertainty, is this one half of Δt? Explain.

D. Choose `MEASURE` and find the time at which the ball's velocity is zero. What is your uncertainty in the reading of v? Call it *dv*. Within your uncertainty, does this velocity vanish at the same time the ball reaches its peak?

III. Modeling Motion with Air Resistance

A. Thinking about Air Resistance

1. Assume that we have two bodies with different masses and drop them. If we ignore air resistance, what are the forces acting on the bodies while they are falling?

2. Write an equation that gives the response of each body to its net force and show that they will have the same accelerations.

3. In the real world, do all falling bodies have the same acceleration? If not, give a real world example.

We might guess that the reason that bodies in the real world do not all fall in the same way is because something else is acting on them other than the earth. Since we know various kinds of contact forces, we might guess that something is in contact with the objects as they fall—the air—and that exerts a force. We'll have to try to make a guess at what that force is.

4. Would you expect the force that air exerts on an object moving through it to depend on the relative velocity between the object and the air or not? Consider your personal experience with wind.

B. Modeling the Force of Air Resistance

Three models of resistive forces are often used:

- Newton drag (force proportional to v^2)
- viscous drag (force proportional to v)
- friction (force independent of the magnitude of v).

We will consider the behavior of a ball that is feeling a drag force. (This is the most realistic for objects the size of a ball.)

1. Choose SET DATA. (If necessary, choose RETURN to return to the menu where you can select SET DATA.) Change the damping coefficient to 0.005 N-s^2/m^2 and press <enter>.

   a. Describe how the position and velocity curves change from the case with no air resistance.

   b. Describe what the changes in the curves mean for how the motion of the ball changes from the case with no resistance.

2. Increase the strength of the air resistance coefficient to 0.01 N-s^2/m^2.

   a. Describe what is happening to the ball at the largest values of t shown.

   b. At these large times is the object accelerating significantly? Use this observation to derive an equation for the value at which the velocity flattens out. This is called the *terminal velocity.*

3. When there is air resistance, does the ball take a longer time to go up or to come down?

4. Explain your answer to question 3 by considering the relative size of the forces on the object as it goes up and as it comes down.

IV. Testing Models of Air Resistance

In this part of the tutorial you will need coffee filters and a meter stick. Coffee filters have the property that they are light enough and have a large enough area that they reach terminal velocity almost immediately. This means that if you drop one at a distance of a meter it floats down smoothly at essentially a constant speed. This will allow us to determine whether they feel an air resistance force that is approximately proportional to v^2 (Newton drag) or to v (viscous drag). (If you don't get to this part of the tutorial during the section, you can easily do this part at home.)

We will write equations that express the condition that the filter is falling at terminal velocity and rearrange it to derive a simple expression for the time it takes to fall a certain distance. From this, we will be able to see how the time of fall depends on the mass (number of nested filters) and on the distance.

A. Suppose a filter of mass m is falling at terminal velocity under the influence of an air resistance force $\mathbf{F}_{fa}$. What is the condition that tells you it will be falling at terminal velocity?

B. Suppose a filter of mass m feels a viscous drag force $\mathbf{F}_{fa} = -b\mathbf{v}$. What will be its terminal velocity?

C. If you let a filter fall a distance S, how long will it take to fall? Express your answer in terms of m, g, b, and S. (Assume it is traveling at terminal velocity the whole time. This is a pretty good approximation.)

D. Suppose you could double the filter's mass without changing its air resistance force. How would the time it takes to fall a distance S change?

You effectively double the filter's mass without changing the air resistance it feels by nesting two filters inside each other. Since only one of them is pushing its way through the air, the air resistance force doesn't change.

E. Find a new distance, D, so that the time it takes the double filter to fall D is the same as the time it takes a single filter to fall a distance S?

F. Take S to be 1 meter. Drop two filters one nested inside each other from the height D at the same time as you drop a single filter from the height S. Do they hit at the same time?

G. Repeat the argument assuming quadratic drag, $F_{fa} = -cv^2$. Now what should D be compared to S? If you use the distances given above, do the filters now hit at the same time? Explain your conclusions.

I. Cart on a Horizontal Spring

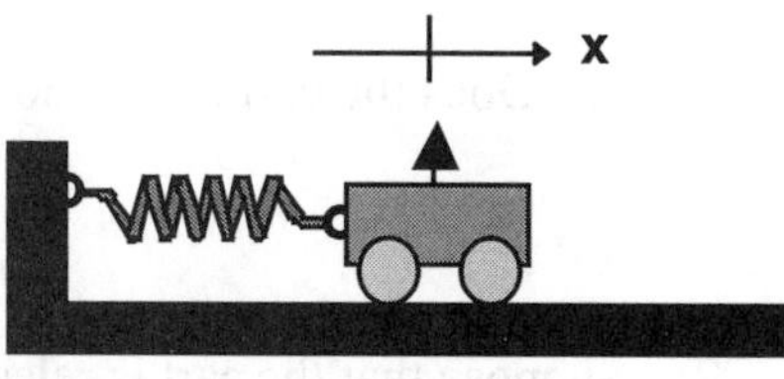

Consider a small cart attached to the end of a spring which is attached to a steel frame, as shown in the diagram.

The cart's displacement, x, is given by the coordinate system shown. The origin is chosen so that when the cart is at rest and the spring is unstretched, its marker is at the origin of the coordinate system ($x = 0$).

A. Consider the cart at the position $x = 0$.

1. Suppose the cart is sitting **at rest**.

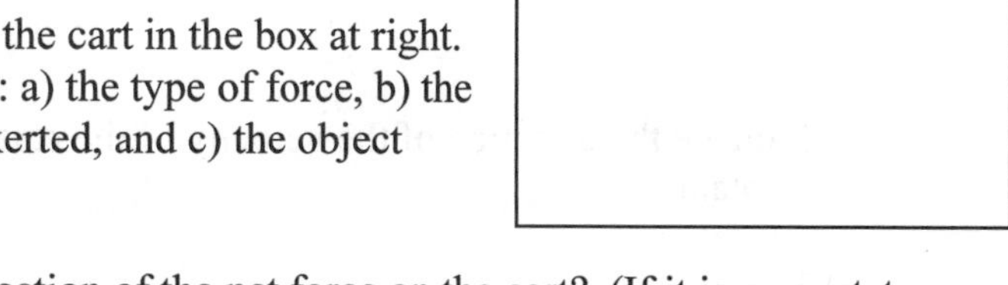

a. Draw a free-body diagram for the cart in the box at right. Label all forces by identifying: a) the type of force, b) the object on which the force is exerted, and c) the object exerting the force.

b. What is the magnitude and direction of the net force on the cart? (If it is zero, state so explicitly.) Explain.

2. Suppose the cart is at the position $x = 0$ but is **moving to the right with a velocity v**.

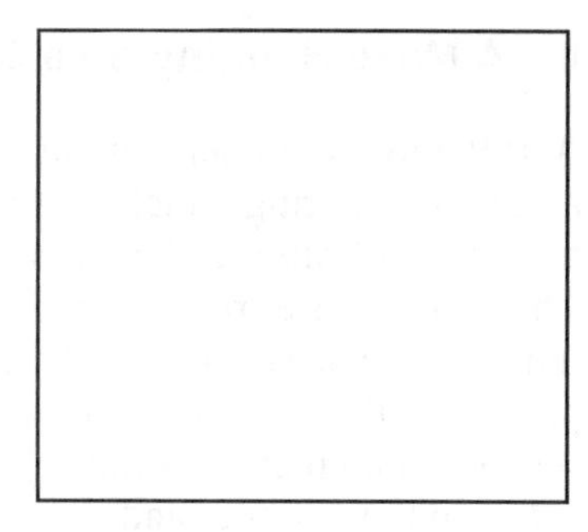

a. Draw a free body diagram of the cart for this instant in time in the box to the right. Label all forces by identifying: a) the type of force, b) the object on which the force is exerted, and c) the object exerting the force.

b. What is the magnitude and direction of the net force on the cart? (If it is zero, state so explicitly.) Explain.

B. Now consider that the cart is located at $x = x_0$ where $x_0 > 0$.

1. Suppose that the cart is held **at rest** by the experimenter's hand.

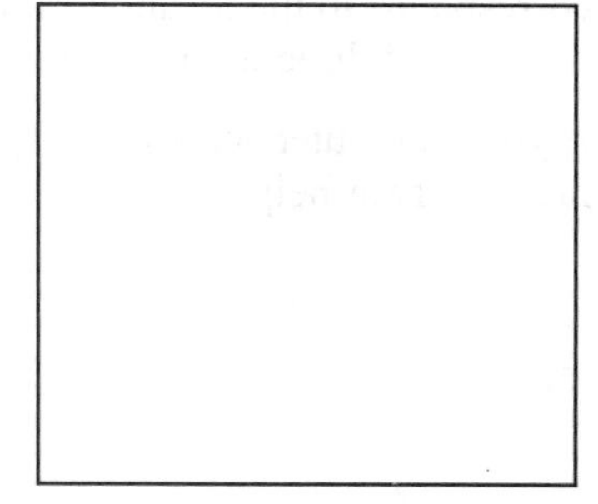

a. Draw a free-body diagram for the cart at this position. Label all forces by identifying: a) the type of force, b) the object on which the force is exerted, and c) the object exerting the force.

b. What is the direction of the net force on the cart?

c. Does the cart have a non-zero acceleration? Explain.

2. Suppose that the cart is released by the experimenter's hand at time $t = t_0$.

a. Draw a free-body diagram for the cart. Label all forces as before. What forces did you add to or remove from your last free body diagram? Explain.

b. Indicate the direction of the velocity of the cart at time t_0. Explain.

c. Indicate the direction of the acceleration of the cart at time t_0. Explain.

II. A Mass Hanging on a Spring

A 100 g iron cylinder is hung from the end of a very light spring which is attached to a heavy steel frame as shown in the diagram. The cylinder is at rest in the position indicated. The spring is hanging from a force probe and the cylinder is above a motion sensor which measures the vertical distance y between the cylinder and the sonic ranger. (See figure.) Output from the motion sensor and force probe are displayed on your computer.

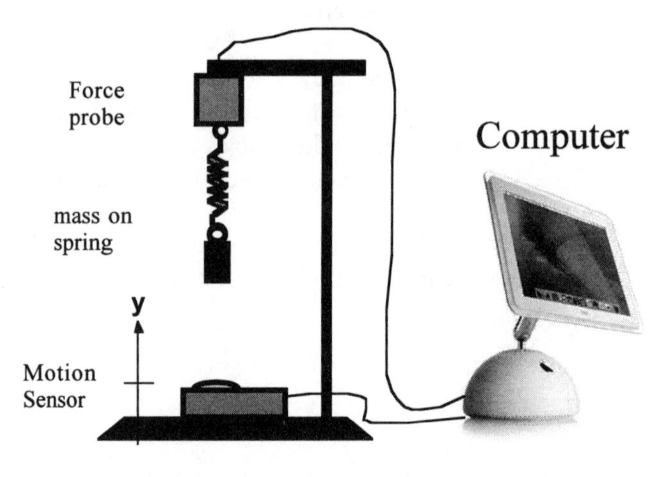

Note that with this arrangement, $y = 0$ corresponds to the location of the ranger and the upward direction is positive.

If your computer screen is not showing the graph windows of the data acquisition program, ask a facilitator for help.

A. Motion

1. On the graph at right, sketch your **prediction** for the shape of the graph y vs t (where t is time) after the cylinder is pulled down slightly and released.

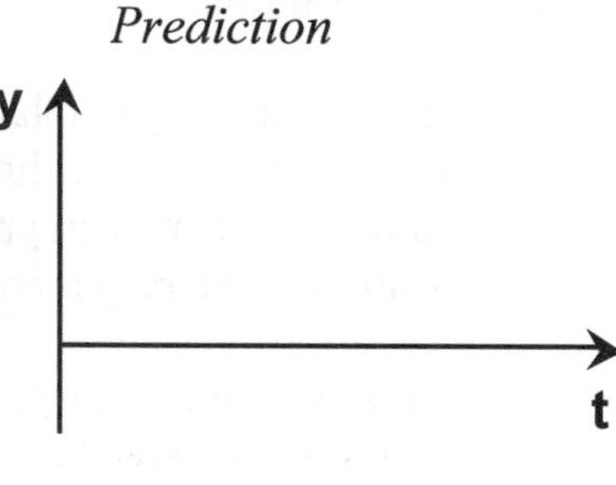

2. **After** sketching your prediction, perform the experiment. On the graph at right, sketch the actual shape of the graph.

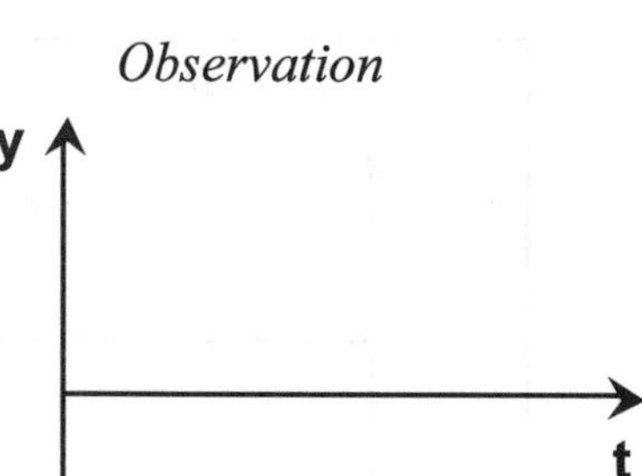

3. Resolve any discrepancies between your prediction and the experiment. Describe your resolutions, if any, in the space below.

B. Spring Constant

1. Set your computer program so that it measures both velocity and force.

   a. Remove the cylinder and spring from the force probe and click on the zero button on the force probe.

   b. Hang the spring from the force probe. Measure the length of your spring when there is no mass attached.

length, no weight	

   c. Place the cylinder back on the spring and allow it to settle down. Measure the length of your spring when the cylinder is attached and at equilibrium.

length with cylinder	

   d. Use the computer to measure the force exerted on the force probe when the cylinder is at rest.

force probe reading	

C. Explain how you can use your measurements to determine the spring constant k. Perform the calculation.

D. Force and Motion

1. On the following graphs, **first** sketch your **prediction** for the shape of the graphs of F vs. t and y vs. t after the cylinder is pulled down slightly and released. Be careful to match up any features in the two graphs that occur at the same time. As before, y = 0 corresponds to the location of the ranger and the upward direction is positive.

2. **After** sketching your prediction, perform the experiment. Sketch the actual shapes of the graphs in the given location.

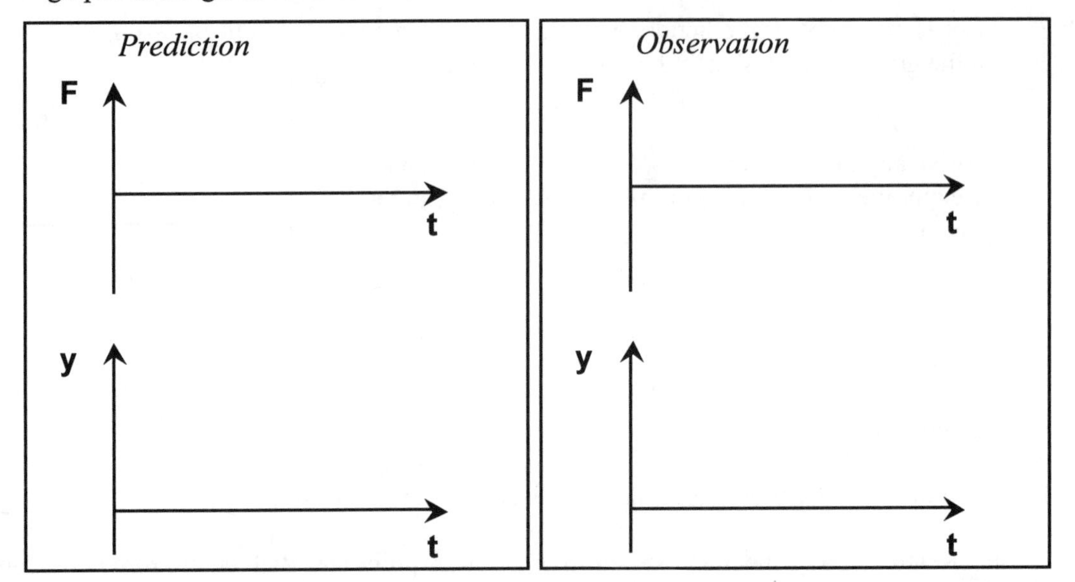

3. Resolve any discrepancies between your prediction and the experiment. Describe the discrepancies, if any, and your resolutions in the space below.

4. Label the equilibrium position of the mass hanging from the spring "y_0."

   a. Write an equation that relates F and y. Explain how you arrived at your answer.

   b. Write an equation that relates y and t. Explain how you arrived at your answer.

   c. Write an equation that relates F and t. Explain how you arrived at your answer.

I. Demonstrations and Videos of Wavepulses on a Spring

A. A facilitator will demonstrate how to create wavepulses on a stretched spring. Observe the motion of the wavepulse and of the spring in each case.

1. Describe what you observed during the demonstrations.

2. A piece of tape has been attached to the spring. How did the motion of the tape compare to the motion of the wavepulse for each type of wavepulse that you observed?

3. The terms *transverse* or *longitudinal* are often used to describe the types of wavepulses you have observed in the demonstration. To what feature of a wavepulse do these terms refer?

For the rest of this tutorial, we will focus on *transverse* wavepulses along the spring.

We have made videos of wavepulses traveling on springs. These videos show a spring like the one you just observed. The video allows you to observe wave motion on a spring more slowly by using the "single frame advance" buttons.

B. On your computer, open ***Shapes.mov***.

1. Describe what you see in the video.

2. Imagine you are holding each of the springs in the video. What hand motion could you use to create wavepulses having these shapes?

C. Open and play the movie ***Amplitud.mov***.

1. Describe what you see in the video.

2. What hand motion could you use to create wavepulses having these shapes?

D. Consider that one of the springs used in questions 1 and 2 is stretched out to a greater length.

1. What physical properties of the spring have been changed by doing this? Explain.

2. Open and play the movie ***Stretched.mov***. In this video, one of the springs has been changed as described in question 1 and the other spring is unchanged. Describe what you see in the video.

E. Summarize your results in this section.

1. Describe how, if at all, you can affect the speed of a wave traveling along a spring.

2. What changes can you make that *do not* affect the speed of the wave? Explain.

3. Are your observations consistent with the equation that describes the speed of a wave on a string, which you learned in class? Resolve any discrepancies.

II. Analysis of a Single Wavepulse

The solid line shown at right indicates the position of a wavepulse traveling along a spring at a time t_0. Each block represents 1 cm. The wavepulse is moving to the right with a speed of 100 cm/sec.

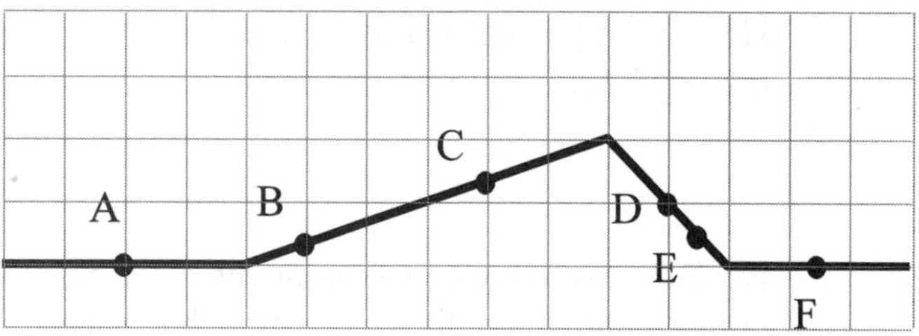

A. In the graph located above, sketch the position of the wavepulse after 0.01 sec has elapsed.

1. How can you use your sketch to find the velocity of a piece of the spring at time t_0 (e.g., the part of the spring labeled D)?

2. Determine the velocity of the piece of spring located at position D at time t_0. Explain.

3. Determine the velocity of the piece of spring located at position C at time t_0. Explain.

B. Consider the instant shown in the diagram above.

1. Draw vectors on the diagram to represent the instantaneous velocity of the pieces of spring labeled A – F at time t_0. Draw your vectors to scale.

2. Compare the direction of motion of the wavepulse and of the spring.

III. The Addition of Wavepulses

A. Open the movie ***SameSide.mov*** on your computer. Play the video. Describe what happens as the two wavepulses meet. Discuss each pulse and the spring.

B. Use the single advance buttons to find the frame showing the moment when the two wavepulses overlap as completely as possible.

1. How could you determine the maximum displacement of the spring at the instant of perfect overlap? Explain.

2. Explain how you can determine the displacement of the spring at locations *other than* the point of maximum displacement at this instant.

C. Find the frame *just before* the moment analyzed above.

1. Describe and sketch the shape of the spring.

2. Account for the shape of the spring between the two peaks. Is your explanation consistent with the explanation you gave in question 2? Resolve any discrepancies.

D. Consider the following situation. Two spring (see upper figure at right). The wavepulses are shown at t = 0 sec. The left pulse moves at a speed of 1 m/sec (=100 cm/sec). Each block represents 1 cm (=0.01 m).

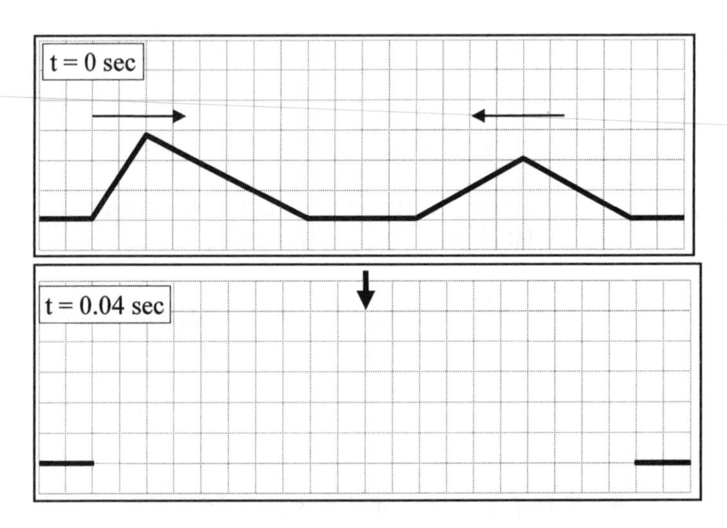

1. What is the speed of the *right* wavepulse? Explain.

2. Consider the spring at time t = 0.04 sec.

   a. Sketch the shape of the spring in the lower figure above. Explain how you arrived at your answer.

b. How did you determine the displacement of the spring at the location of the gridline indicated by the arrow? Explain.

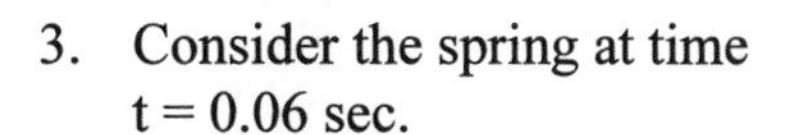

3. Consider the spring at time t = 0.06 sec.

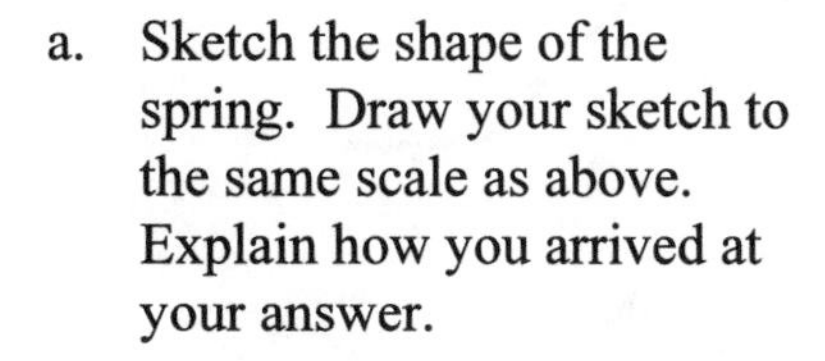

a. Sketch the shape of the spring. Draw your sketch to the same scale as above. Explain how you arrived at your answer.

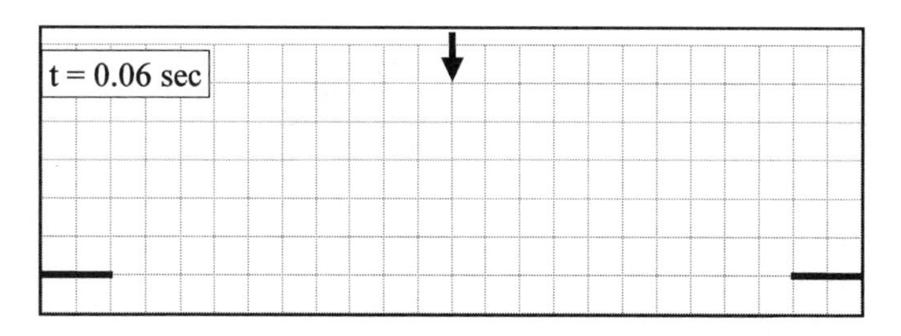

b. How can you determine the displacement of the spring at *any location* and *any time* when two wavepulses overlap. Explain you reasoning.

E. Summarize your results.

1. Explain how the answers you have given in this section (specifically in question 3.b) are consistent with the *Principle of Superposition.*

2. Are the situations in this section of the tutorial examples of constructive or destructive superposition? Explain.

II. Destructive Superposition

Two wavepulses on opposite sides of a spring move toward each other at 100 cm/s. The diagrams below show the wavepulse locations at three successive instants, 0.01 s apart. (In the last diagram, the individual wavepulses are shown dashed.) Each square represents 1 cm.

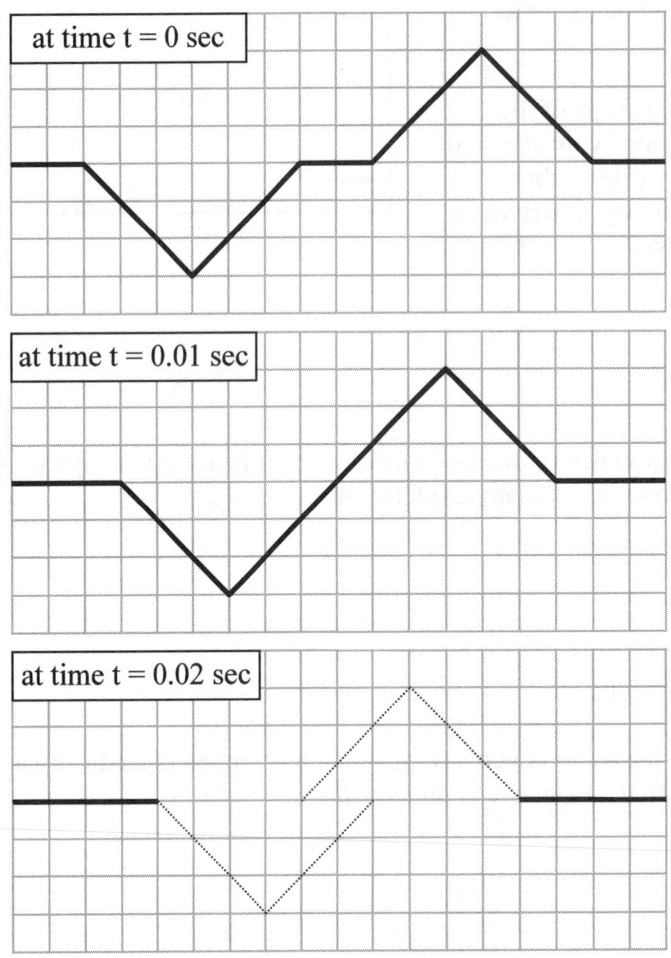

A. Use the principle of superposition to find the shape of the spring at t = 0.02 sec. Draw it in the graph above. Make sure all of your group agrees on how you arrived at your answer before continuing.

B. Three further time steps are reproduced on the next page of this tutorial. Each person in your group should draw what the spring will look like for ONE of the times shown from 0.03 sec to 0.05 sec. Draw the shape of the **spring** for each of the instants shown. After constructing your diagram, discuss your results with the rest of your group until you all agree what the spring should look like at each instant.

   1. Are your graphs consistent with your explanation in E.1? Resolve any discrepancies.

at time t = 0.03 sec

at time t = 0.04 sec

at time t = 0.05 sec

2. Consider the spring at time t = 0.04 sec. Describe the shape of the spring.

C. On your computer, open and play the movie ***Opposite.mov***.

1. Describe what happens in the movie.

2. Is there ever a moment when the spring is perfectly straight? Explain how your observation is consistent with the results of the sketches you made on the large diagrams.

3. Do the wavepulses bounce off each other or pass through each other? Explain. (Is your response consistent with your observation of constructive superposition?)

D. Compare the movie ***Opposite.mov*** to the sketches on the previous page.

1. How can you account for the continued propagation of the wavepulses after the time t = 0.04 sec?

2. Sketch a graph of the velocity as a function of position for the spring at time t = 0.4 sec in the space below. Explain how you arrived at your answer.

at time t = 0.04 sec

I. Describing the Movement of a Wavepulse

A. At t = 0 sec, the displacement of a spring from its equilibrium position can be written as a function of x, $y(x) = \dfrac{50cm}{(x/b)^2 + 1}$, where b = 20 cm.

1. Sketch the shape of the spring at t = 0 sec in the graph below. Use a scale where each block represents 10 cm on a side.

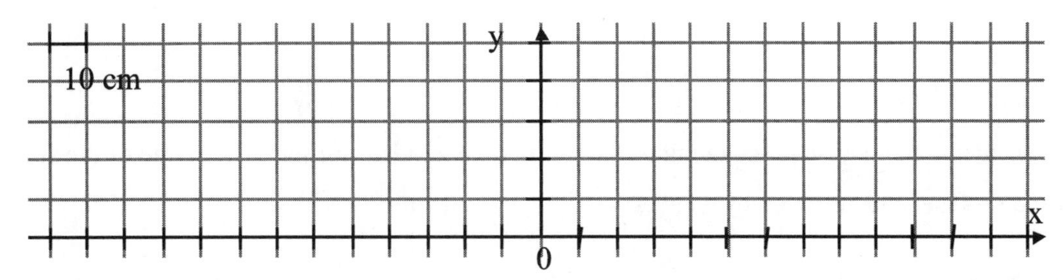

2. The digitized picture below shows a wavepulse propagating to the right on a smooth surface. The wavepulse loses nearly no energy as it travels along the segment of spring shown in the picture. The wavepulse can be roughly described by an equation of the form $y(x) = \dfrac{A}{(x/b)^2 + 1}$.

   Using the indicated scale, find approximate values for A and b.

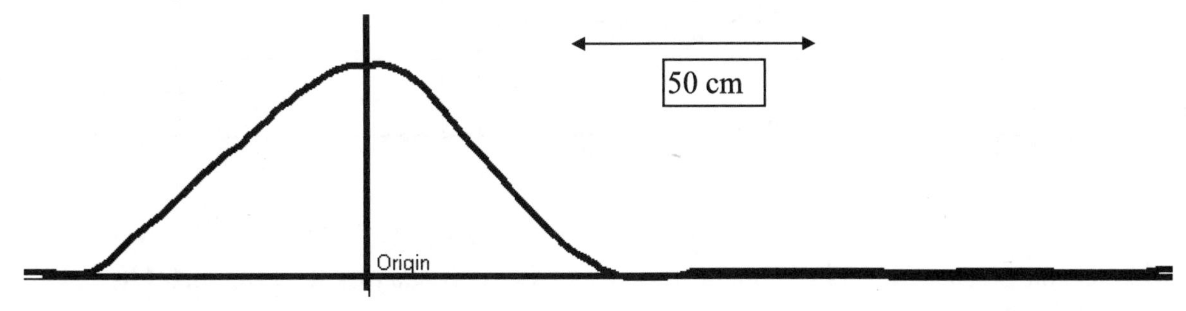

Explain how you arrived at your answer.

3. Predict the maximum amplitude of the wavepulse after its peak has moved a distance 3b to the right. Explain how you arrived at your answer.

4. On your computer, open the movie ***Pulse.mov***. Play the video. Compare what you see in the video to your answer to question 3. Resolve any discrepancies.

5. Describe what the symbol "x" represents in the equation on the previous page. Explain.

B. Consider a wavepulse propagating to the right along an ideal spring. The shape of the spring at time t = 0 seconds is given by $y(x) = \dfrac{50cm}{(x/b)^2 + 1}$, b = 20 cm.

1. In the space below, sketch the shape of the spring after the wavepulse has moved so that its peak is at x = 70 cm. Compare this graph to the graph you sketched in question 1 on the previous page.

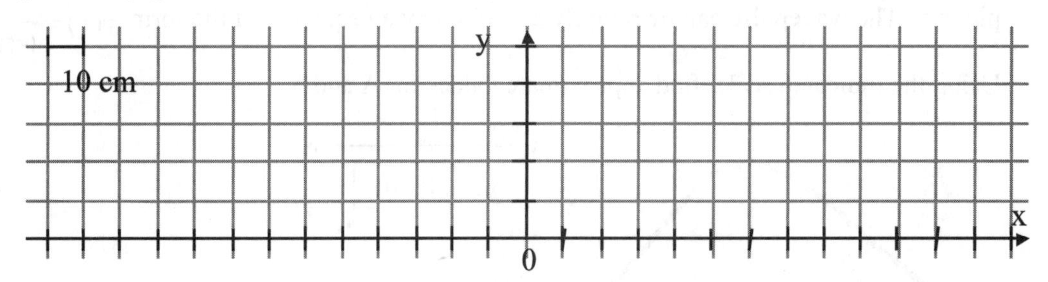

2. On the graph above, draw a coordinate system with its vertical axis at 70 cm. Label its horizontal axis with the variable *s*.

3. Write a formula y(s) that describes the displacement of any piece of the spring (when the peak of the wavepulse is located at 70 cm) *as a function of s*. Explain.

4. Write an equation for *s* as a function of *x*.

5. Write a formula that describes the shape of the wavepulse at the time it is centered at 70 cm *as a function of x*. Explain.

6. Consider that the wavepulse had moved an arbitrary distance x_0. How would your formula change? Explain.

7. Suppose the wavepulse moved a distance x_0 at a speed v in a time t_0. Write an equation for x_0 in terms of v and t_0.

8. How could you use this information to find the displacement of any piece of the spring at time t_0? Write an equation that would let you do this. Explain.

9. Write an equation that describes the displacement of any piece of the spring at *any* time. Explain.

10. Describe how you would find the displacement of any piece of the spring at any time.

11. Would your equation in question 8 be correct if the spring were *not* ideal, or if there were friction between the spring and the ground? Explain.

II. Measuring the Shape of the Wavepulse

A. Consider a different wavepulse propagating along a long, taut spring. The diagram below shows the shape of the wavepulse at time t = 0 sec. Suppose the displacement of the spring at various values of x is given by $y(x) = Ae^{-\left(x/b\right)^2}$. The wavepulse moves with a velocity v to the right.

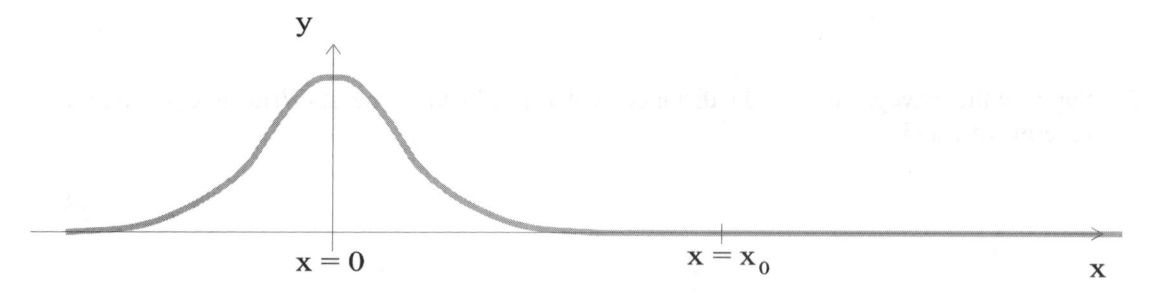

1. On the diagram above, sketch the shape of the spring after it has traveled a distance x_0, where x_0 is shown in the figure. Explain why you sketched the shape you did.

2. Write an equation that describes the displacement of the spring as a function of x and t for the instant of time that you have sketched. Explain how you determined your answer.

A piece of tape is attached to the spring at position $x = x_0$.

3. In the space below, *qualitatively* sketch the velocity of the piece of tape as a function of time. Explain how you arrived at your answer.

4. In the space below, *qualitatively* sketch the acceleration of the piece of tape as a function of time. Explain how you arrived at your answer.

B. On your computer, play ***pulse.vpt*** to show a video of a single wavepulse traveling along a spring. Suppose the wavepulse in the video can be described by an equation like the one you wrote in question 2 on the previous page.

1. What effect would changing the parameters "A", "b", and "v" in the equation (from page 50) have on the wavepulse?

2. Find numerical values for each parameter in the equation. Show all work.

3. Use your equation to find the displacement of the spring at position x = 125 cm at time t = 0.1 sec. Show all work.

4. Advance the video 0.1 sec beyond the instant of time where its peak is located at the origin. Find the displacement of the spring at position x = 125 cm and time t = 0.1 sec.

5. Compare your answer to question 4 with your answer to question 3. Resolve any discrepancies.

I. Introduction

In this tutorial, we will consider sound waves. Experiments show that sound waves travel at about 340 m/s through air at room temperature.

A. Consider a flame placed in front of a speaker as shown in the figure at right. No wind is blowing.

1. The speaker plays a note at a constant pitch. Explain how, if at all, the sound produced by the speaker affects the flame. If the sound does not affect the flame, state that explicitly.

2. Open the movie ***Sound Movie.mov*** on your desktop. Play the movie.

   a. Describe your observations. Do your observations agree with your earlier predictions?

   b. How can you account for the flame's motion? Explain your reasoning.

B. Consider a dust particle in front of a silent loudspeaker. The dust particle is in the same location as the original location of the flame. The speaker is turned on and plays a note with a constant frequency, f. How, if at all, does this affect the motion of the dust particle? Explain.

II. Motion of a Single Flame

We have used a program to analyze the motion of the flame in the video. By measuring the position of a point on the left edge of the candle flame, we can analyze the motion of the flame.

A. Data from our measurements are shown to the right.

Time (s)	x (mm)	y (mm)
0	0	0
0.033	2.8	0
0.067	2.1	0
0.1	0	0
0.13	-3.5	0
0.17	-1.4	0
0.2	1.4	0
0.23	2.8	0
0.27	2.1	0
0.3	-2.1	0
0.33	-3.5	0
0.37	-1.4	0
0.4	2.1	0
0.43	2.8	0
0.47	-2.1	0
0.5	-3.5	0
0.53	-1.4	0
0.57	2.1	0
0.6	3.5	0
0.63	2.1	0
0.67	-2.1	0
0.7	-2.1	0

1. In the space below, plot the position of the flame. Label axes clearly!

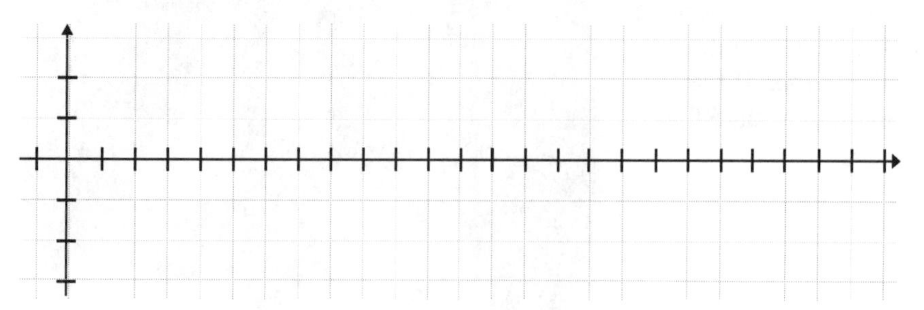

2. What does each axis on your graph represent? Explain.

B. Describe the graph you have sketched in general terms.

1. Describe the shape of your graph. (It may help to sketch a continuous curve on the basis of your data points.)

2. Find the period with which the flame and speaker oscillate. Explain how you obtained your answer.

3. How many oscillations are there in a 1 second time interval. Explain how you obtained your answer.

III. Motion of Many Flames

A. Consider two candles sitting 25 cm apart in front of a loudspeaker oscillating at 680 Hz. A clock is started at an arbitrary time. At time t = 0 seconds, the first flame is perfectly vertical and moving away from the speaker. Its maximum displacement from equilibrium is 5 mm.

1. In the graph below, sketch the displacement of the first flame from equilibrium at different times. Label axes clearly. Explain how you arrived at your answer.

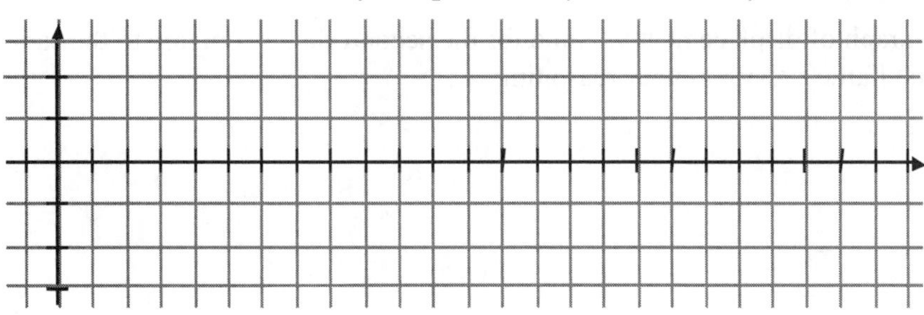

2. Compare the motion of the second flame to the motion of the first flame. Discuss both frequency and how they are moving at a given instant in time.

3. In the graph below, sketch the displacement of the second flame for the entire time considered. Describe in words how you determined the shape of the sketch.

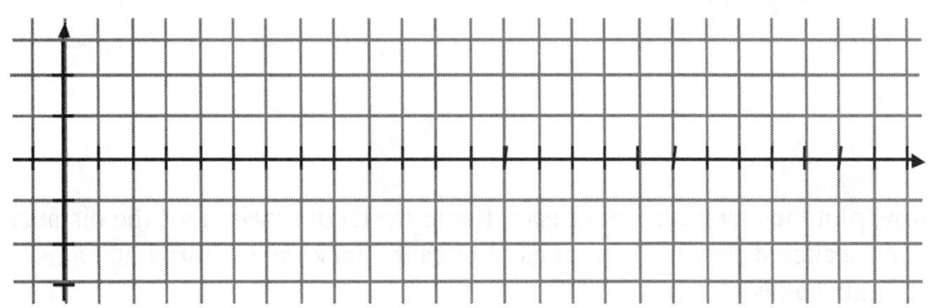

B. Consider an arrangement of candles placed 12.5 cm apart, and the first is located 12.5 cm from the speaker. The speaker plays a note at 680 Hz. At an arbitrary time after all the flames are in motion, a clock is started. At time t = 0 seconds, the first flame is at its maximum displacement of 5 mm from its upright position, away from the loudspeaker.

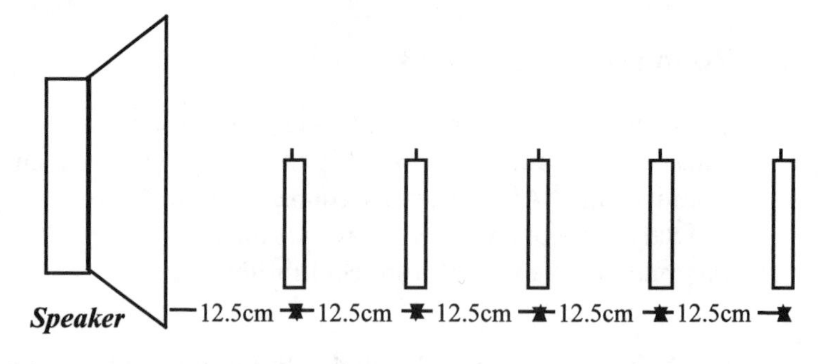

1. How would the graph of displacement vs. time for the second flame compare to the same graph for the first flame? Explain your reasoning.

2. The displacement of flame 1 at time t = 0 seconds is at a maximum and is shown at right. (The displacement is exaggerated for easy viewing). Sketch the displacement of each of the other flames at time t = 0 seconds in the diagram. Explain how you arrived at your answer.

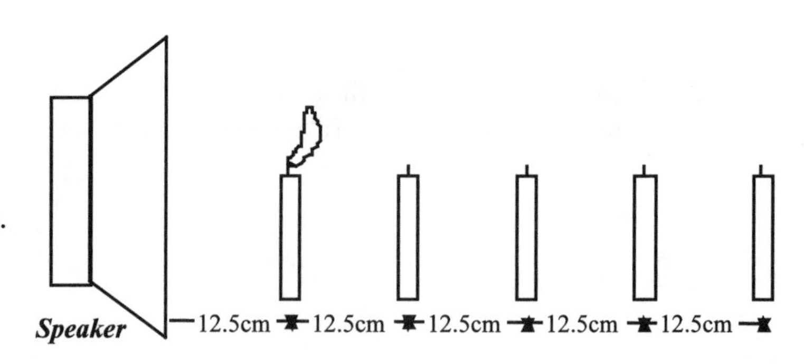

3. In the graph below, plot the displacement of each flame from equilibrium vs. the distance of the flame from the speaker at time t = 0 seconds. Let each block on the horizontal axis represent 2.5 cm. Label axes clearly.

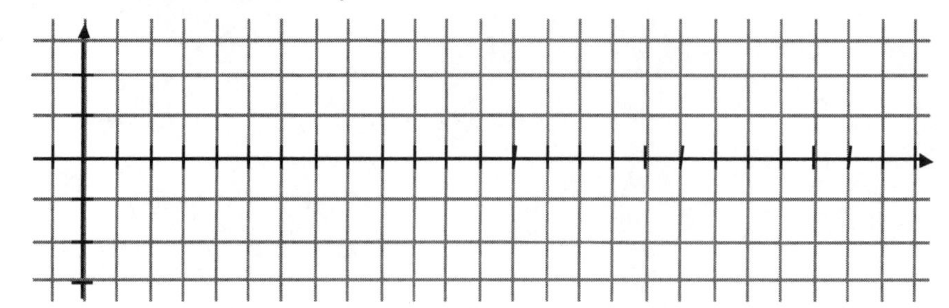

4. Consider a very large number of flames placed from 5 cm to 60 cm away from the speaker. On the graph on the previous page, plot the displacement of each of these flames at time $t = 0$ seconds as a function of distance from the speaker. Explain how you arrived at your answer.

5. Describe the shape of the graph you have sketched.

6. How can you use the graph you have sketched to find the wavelength of the sound produced by the loudspeaker? Explain.

7. Find the wavelength of the sound produced by the speaker in the movie ***sound.mov***.

IV. Giving Multiple Interpretations of a Situation

A. Consider the information given by different graphs.

1. How, if at all, could you use the graph on page 54 (instead of the graph on page 56) to find the wavelength of the sound wave?

2. How, if at all, could you use the graph on page 56 (instead of the graph on page 54) to find the frequency of the sound wave? the period of the sound wave?

3. How, if at all, could you use the graphs on pages 54 and 56 to find the amplitude of the sound wave?

B. Consider the candles shown in the sketch on page 56 at different times after t = 0 sec.

1. In the figure to the right, sketch the displacement of each of the flames at time t = T/4 seconds (where T is the period). Explain how you arrived at your answer.

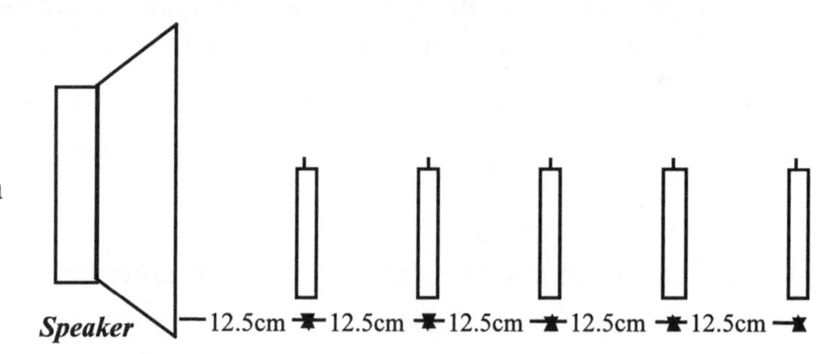

2. In the figure to the right, sketch the displacement of each of the flames at time t = T/2 seconds. Explain how you arrived at your answer.

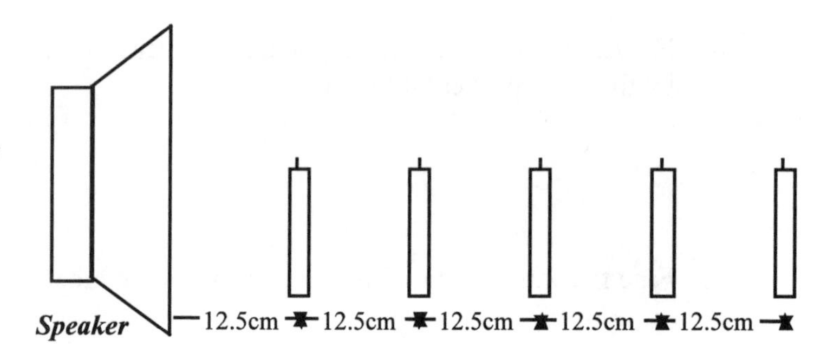

3. In the figure to the right, sketch the displacement of each of the flames at time t = 3T/4 seconds. Explain how you arrived at your answer.

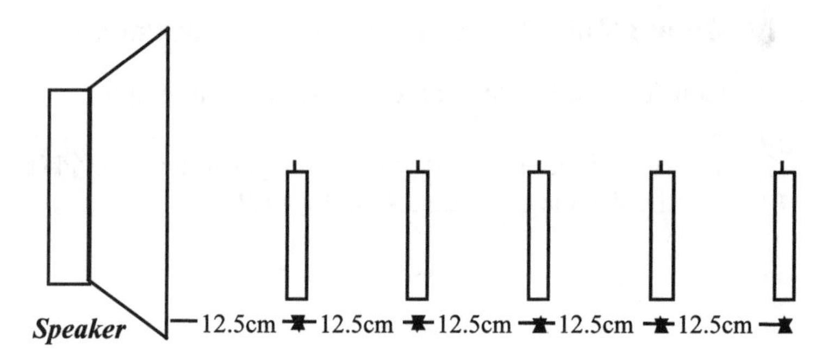

4. In the figure to the right, sketch the displacement of each of the flames at time t = T seconds. Explain how you arrived at your answer.

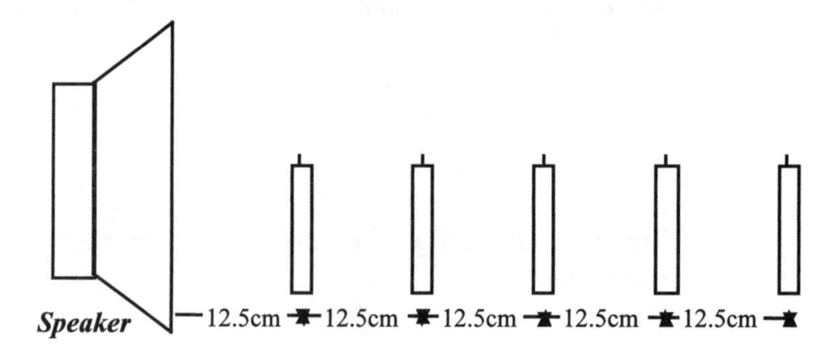

I. Introduction

In this tutorial, we will consider sound waves. Experiments show that sound waves travel at about 340 m/s through air at room temperature.

A. Consider a flame placed in front of a speaker as shown in the figure at right. No wind is blowing.

1. The speaker plays a note at a constant pitch. Explain how, if at all, the sound produced by the speaker affects the flame. If the sound does not affect the flame, state that explicitly.

2. Open the movie ***Sound Movie.mov*** on your desktop. Play the movie.

   a. Describe your observations. Do your observations agree with your earlier predictions?

   b. How can you account for the flame's motion? Explain your reasoning.

B. Consider a dust particle in front of a silent loudspeaker. The dust particle is in the same location as the original location of the flame. The speaker is turned on and plays a note with a constant frequency, f. How, if at all, does this affect the motion of the dust particle? Explain.

II. Motion of a Single Flame

The program VideoPoint lets you analyze the motion of the flame by giving information about the positions of points on the screen at different times.

A. On the desktop, open the file ***Sound Data File.vpt***. This lets you analyze the position of points on the video screen for the movie. Play the video using the single advance buttons.

1. What is the red cross on the video screen measuring? Explain.

2. The "Table" window of ***Sound Data File.vpt*** includes data already obtained. These data are also shown to the right. What does this data represent?

Time (s)	x (mm)	y (mm)
0	0	0
0.033	2.8	0
0.067	2.1	0
0.1	0	0
0.13	-3.5	0
0.17	-1.4	0
0.2	1.4	0
0.23	2.8	0
0.27	2.1	0
0.3	-2.1	0
0.33	-3.5	0
0.37	-1.4	0
0.4	2.1	0
0.43	2.8	0
0.47	-2.1	0
0.5	-3.5	0
0.53	-1.4	0
0.57	2.1	0
0.6	3.5	0
0.63	2.1	0
0.67	-2.1	0
0.7	-2.1	0

3. In the space below, plot the position of the flame.

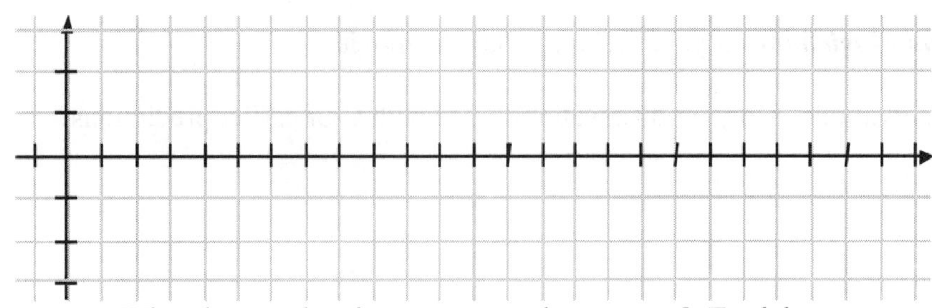

What does each axis on your graph represent? Explain.

B. Describe the graph you have sketched in general terms.

1. Describe the shape of your graph. (It may help to sketch a continuous curve on the basis of your data points.)

2. Find the period with which the flame and speaker oscillate. Explain how you obtained your answer.

3. How many oscillations are there in a 1 second time interval. Explain how you obtained your answer.

III. Motion of Many Flames

A. Consider two candles sitting 25 cm apart in front of a loudspeaker oscillating at 680 Hz. A clock is started at an arbitrary time. At time t = 0 seconds, the first flame is perfectly vertical and moving away from the speaker. Its maximum displacement from equilibrium is 5 mm.

1. In the graph below, sketch the displacement of the first flame from equilibrium at different times. Label axes clearly. Explain how you arrived at your answer.

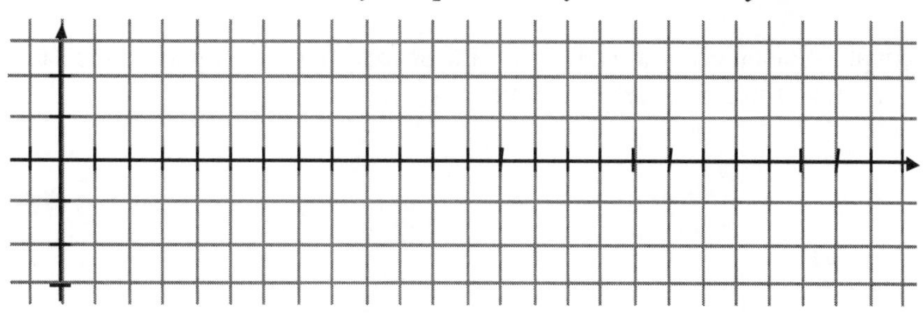

2. Compare the motion of the second flame to the motion of the first flame. Discuss both frequency and how they are moving at a given instant in time.

3. In the graph below, sketch the displacement of the second flame for the entire time considered. Describe in words how you determined the shape of the sketch.

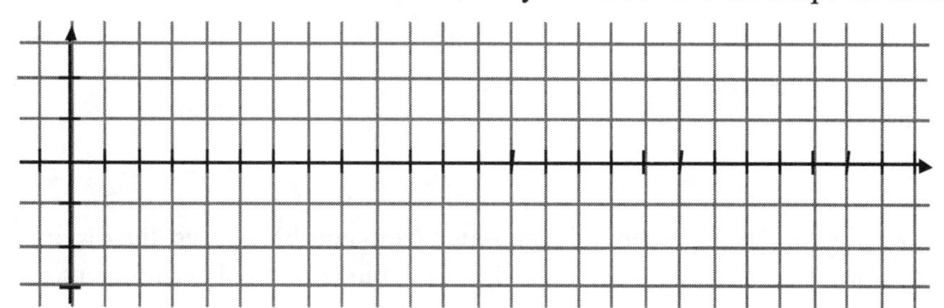

B. Consider an arrangement of candles placed 12.5 cm apart, and the first is located 12.5 cm from the speaker. The speaker plays a note at 680 Hz. At an arbitrary time after all the flames are in motion, a clock is started. At time t = 0 seconds, the first flame is at its maximum displacement of 5 mm from its upright position, away from the loudspeaker.

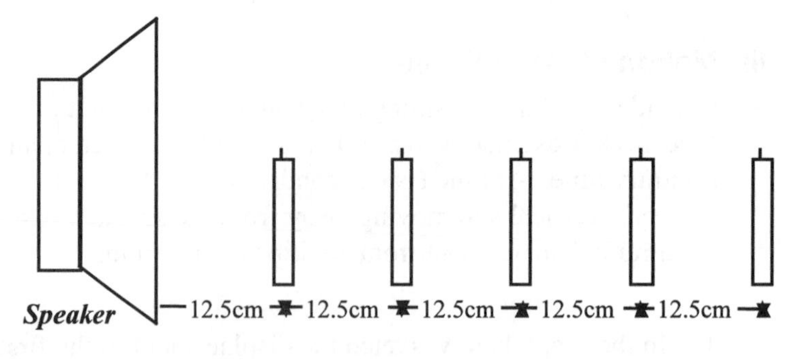

1. How would the graph of displacement vs. time for the second flame compare to the same graph for the first flame? Explain your reasoning.

2. The displacement of flame 1 at time t = 0 seconds is at a maximum and is shown at right. (The displacement is exaggerated for easy viewing). Sketch the displacement of each of the other flames at time t = 0 seconds in the diagram. Explain how you arrived at your answer.

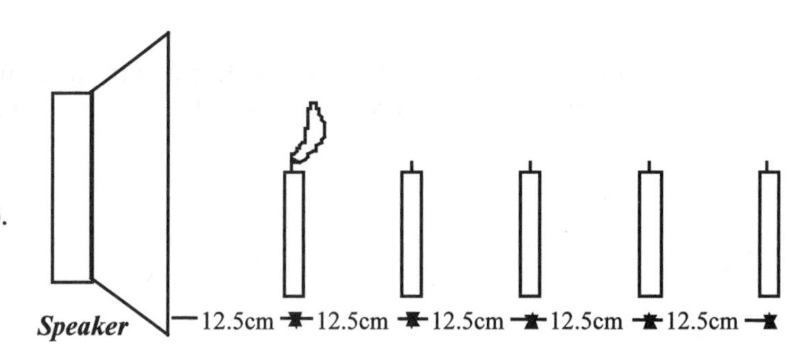

3. In the graph below, plot the displacement of each flame from equilibrium vs. the distance of the flame from the speaker at time t = 0 seconds. Let each block on the horizontal axis represent 2.5 cm. Label axes clearly.

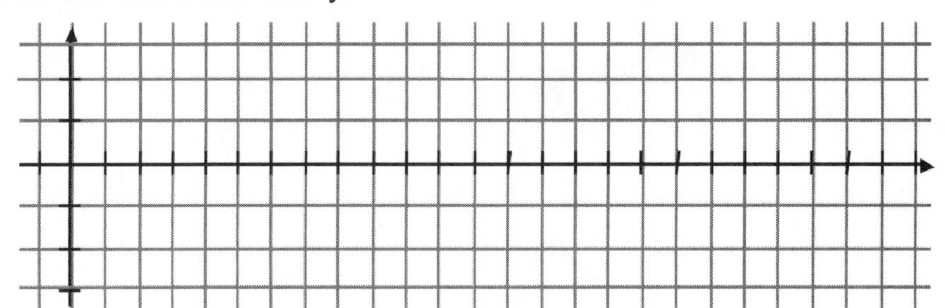

4. Consider a very large number of flames placed from 5 cm to 60 cm away from the speaker. On the graph on the previous page, plot the displacement of each of these flames at time t = 0 seconds as a function of distance from the speaker. Explain how you arrived at your answer.

5. Describe the shape of the graph you have sketched.

6. How can you use the graph you have sketched to find the wavelength of the sound produced by the loudspeaker? Explain.

7. Find the wavelength of the sound produced by the speaker in the movie ***sound.mov***.

IV. Giving Multiple Interpretations of a Situation

A. Consider the information given by different graphs.

1. How, if at all, could you use the graph on page 60 (instead of the graph on page 62) to find the wavelength of the sound wave?

2. How, if at all, could you use the graph on page 62 (instead of the graph on page 60) to find the frequency of the sound wave? the period of the sound wave?

3. How, if at all, could you use the graphs on pages 60 and 62 to find the amplitude of the sound wave?

B. Consider the candles shown in the sketch on page 62 at different times after t = 0 sec.

1. In the figure to the right, sketch the displacement of each of the flames at time t = T/4 seconds (where T is the period). Explain how you arrived at your answer.

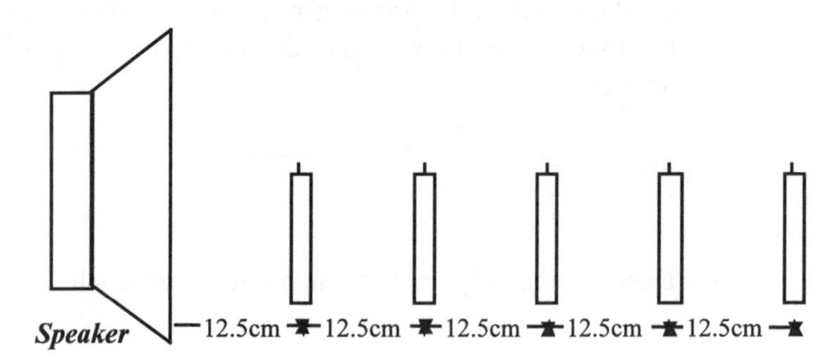

2. In the figure to the right, sketch the displacement of each of the flames at time t = T/2 seconds. Explain how you arrived at your answer.

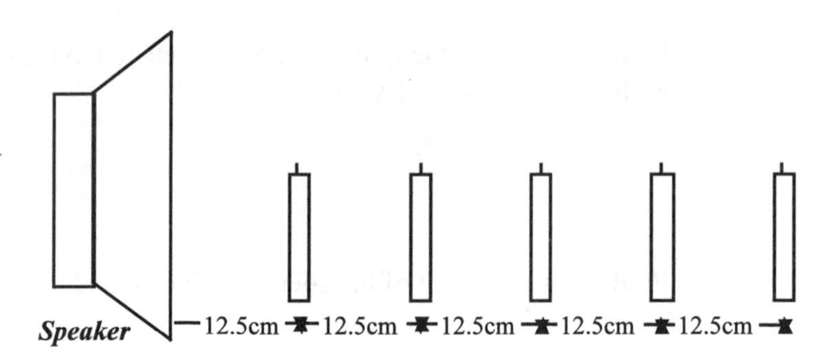

3. In the figure to the right, sketch the displacement of each of the flames at time t = 3T/4 seconds. Explain how you arrived at your answer.

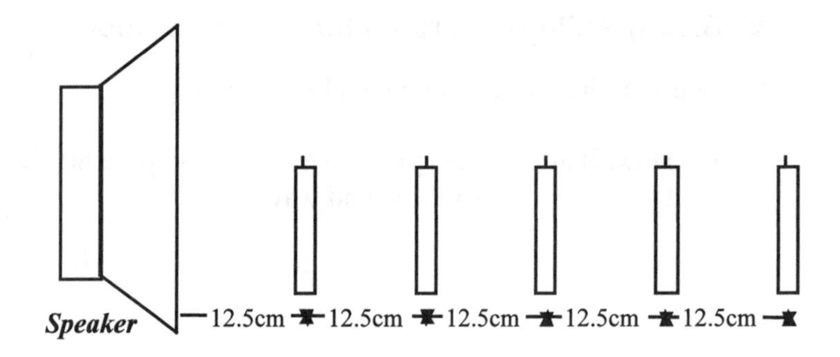

4. In the figure to the right, sketch the displacement of each of the flames at time t = T seconds. Explain how you arrived at your answer.

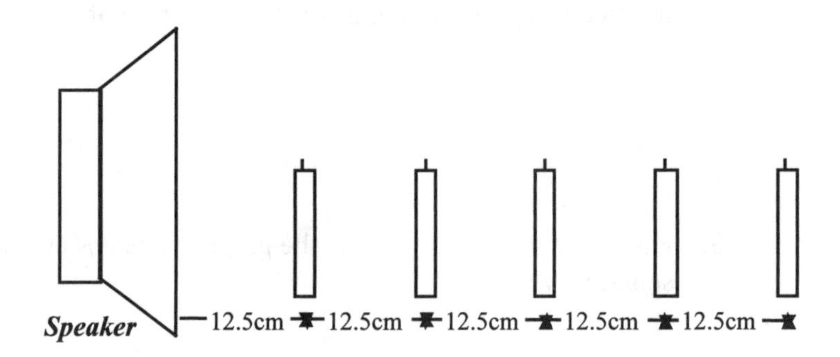

In this tutorial we will use thermometers connected to the computer to measure and display the temperature of objects. We will *define* temperature as the reading on the thermometer.

A computer program will collect and display temperature data gathered from two temperature probes.

I. Mixing Equal Amounts of Water

A. Predicting the Final Temperature

Before you use the program, answer the following questions. Consider two cups that contain equal amounts of water at different temperatures.

1. Describe in your own words what you think will happen when they are mixed together. Explain how you arrived at your answer.

2. Suppose one cup of water is at a temperature T_1 and the other at a higher temperature T_2. How will the temperature of each change when they are mixed together? Explain.

3. The program plots the temperature measured by each probe as a function of time. If you measured the temperatures in the experiment described above, sketch what you think the plot would look like on the graph below.

Explain how you arrived at your answer.

4. Describe the direction of temperature change (positive or negative) that will occur. In your experience, is it ever possible that when a hot and cold object are put together that the hot object becomes hotter and the cold object becomes colder? Explain.

B. Observing the Temperature Change

Now perform the experiment directly. Fill the small styrofoam cup a third full with hot water provided by the TA. Using the balance scale, measure out an equal amount of room temperature water from the bucket in the room and put it into the other small styrofoam cup. Place the cup with the room temperature water in the plastic bowl. Place a temperature probe in each cup and make sure the temperature the probe is reading (shown at the bottom of the computer window) has stabilized. Then start the program.

Remember that it takes several seconds for the computer to respond and begin recording temperatures.

Once the program has started recording the temperatures in each cup, transfer the water and the probe from the cup with the hot water to the cup with the room temperature water in the plastic bowl. Stir the mixed water.

1. Describe the result. How does the temperature change recorded by one probe compare with the temperature change recorded by the second?

   Explain in your own words why you think this occurs.

2. At an instant a few seconds after you mixed the two cups of water together, the thermometers showed *different* temperatures. With the water all mixed together, how did each thermometer recognize its "own" water? Why weren't the two temperature readings the same as soon as the water was mixed? Explain.

II. Mixing Different Amounts of Water

A. Predicting the Final Temperature

In the next experiment, we will mix *different* amounts of water at *different* temperatures.

1. Describe in your own words what you think will happen to one cup of hot water and two cups of cold water when mixed together. How will the temperature of each change? Be as specific as possible.

2. On the graph below, sketch what you think the plot of the temperatures will look like.

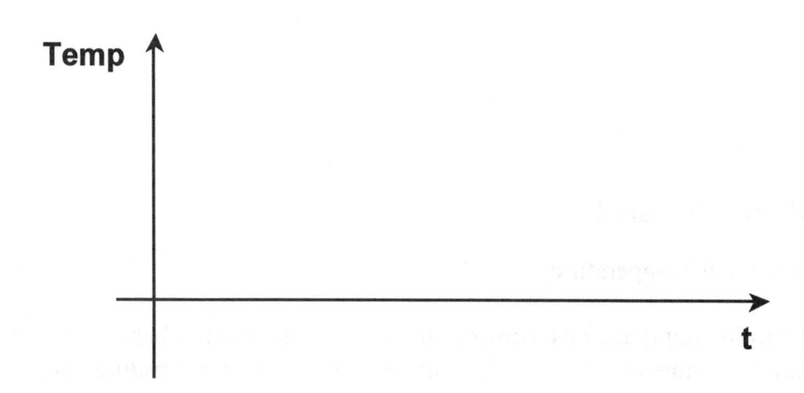

B. Observing the Temperature Change

Now perform the experiment directly. Obtain about a half cup of hot water in a small styrofoam cup from the TA in the front of the room. Using the balance scale, measure out two equal amounts of colder water and combine these two in the large styrofoam cup. Position the temperature probes in the hot and cold water and start the program. Once the program has started recording, transfer the hot water and its probe into the large styrofoam cup with the cold water.

1. Describe the results as shown on the screen. Account for any discrepancies with your prediction.

2. Measure the change in temperature of the hot water. Measure the change in temperature of the cold water. Write your values in the space below.

3. Consider a situation where you had half of the original amount of cold water. Compare its temperature change with the temperature change of the entire original amount of cold water. Explain.

C. Consider two cups of water at different temperatures. The mass of the cold water is m_1 and the mass of the hot water is m_2. Suppose the two cups are mixed. Call the temperature change of the hot and cold water ΔT_1 and ΔT_2, respectively.

Use your earlier results to help you find the relationship between m_1, m_2, ΔT_1, and ΔT_2. Explain how you arrived at your answer.

III. Mixing Different Materials

A. Predicting the Final Temperature

In answering the following questions, consider the thermal interaction between two different materials that have the same mass, e.g. a piece of copper at room temperature and a cup of hot water.

1. What do you think will happen to the two objects when combined? How will the temperature of each change? Explain how you arrived at your answer.

2. In the graph below, sketch what you think the graph of temperature vs. time for each quantity is going to look like.

3. What is different about this experiment from the previous ones? What is the same?

B. Observing the Temperature Change

Perform the experiment directly. Take one copper cylinder and, using the balance scale, put an equal mass of room temperature water into the large styrofoam cup. Place one temperature probe in the cup of room temperature water and one into the cylindrical hole in the top of the copper cylinder and start the program.

Once the program has started recording the temperatures, pour the water and transfer its temperature probe to the cup containing the copper. Make sure the temperature probe touching the copper does not fall off.

1. Describe the results as shown on the screen. How does the temperature change recorded by one probe compare with the temperature change recorded by the second? Is this what you predicted? Account for any discrepancies with your prediction.

2. Does the relationship you determined in II.C hold in this situation? Explain why you think it does or does not.

C. Suppose the copper is again heated to the same temperature as before. Predict the amount of water that would be needed such that the temperature change of the water equals the temperature change of the copper. Explain how you came up with your prediction.

Conduct the experiment. Account for any discrepancies with your prediction.

I. Introduction

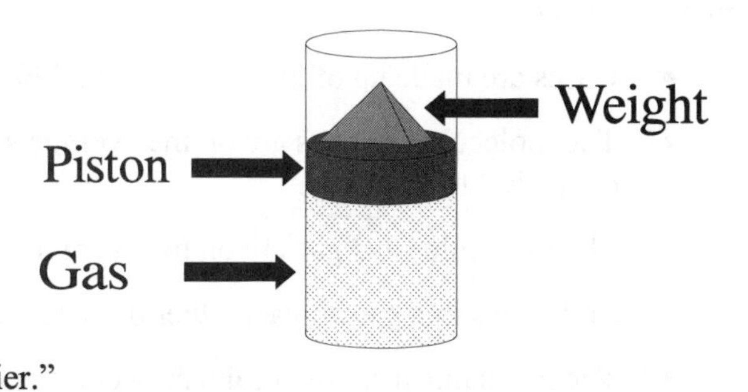

One of the most striking properties of a gas is its compressibility—what Robert Boyle called "the spring of the air." Consider an airtight cylinder having nothing in it (but air) and capped by a smoothly sliding airtight cylinder (a piston). The "empty space" in the cylinder (the air) will support the weight of the piston and more. As more weight is put on top, the gas is compressed more and more but becomes "springier and springier."

When this phenomenon was observed at the end of the 18th century, many leading scientists made models for the structure of the gas responsible.

A. Come up with at least one model for the gas that would account for these observations.

B. If you have not already done so, compare your model to the model of your group members. Describe your group model in the space below.

The model that has proved most effective in explaining a large variety of phenomena was developed by Clausius, Maxwell, and others in the middle of the 19th century. The primary characteristics of this model are:

- Gases are made up of molecules—small bits of matter.
- The molecules in a gas are on the average separated from each other by distances large compared to their size.
- The molecules in a gas are on the average moving very fast.
- Molecules in a gas hit each other occasionally.
- The mechanical energy of the molecules is conserved, both in-between collisions and during a collision.

A number of these properties are unusual and don't correspond to our everyday experience.

C. Which of the above properties seem to you to be the strangest? Why?

D. Predict the motion of a *single* molecule in a gas. How can you account for your prediction?

II. Using a Program to Model a Gas

On your computer, you will need to run the program THERMO on the desktop. When the program's title screen appears, press `<spacebar>` to start the program.

The program's display screen will appear. It looks something like this.

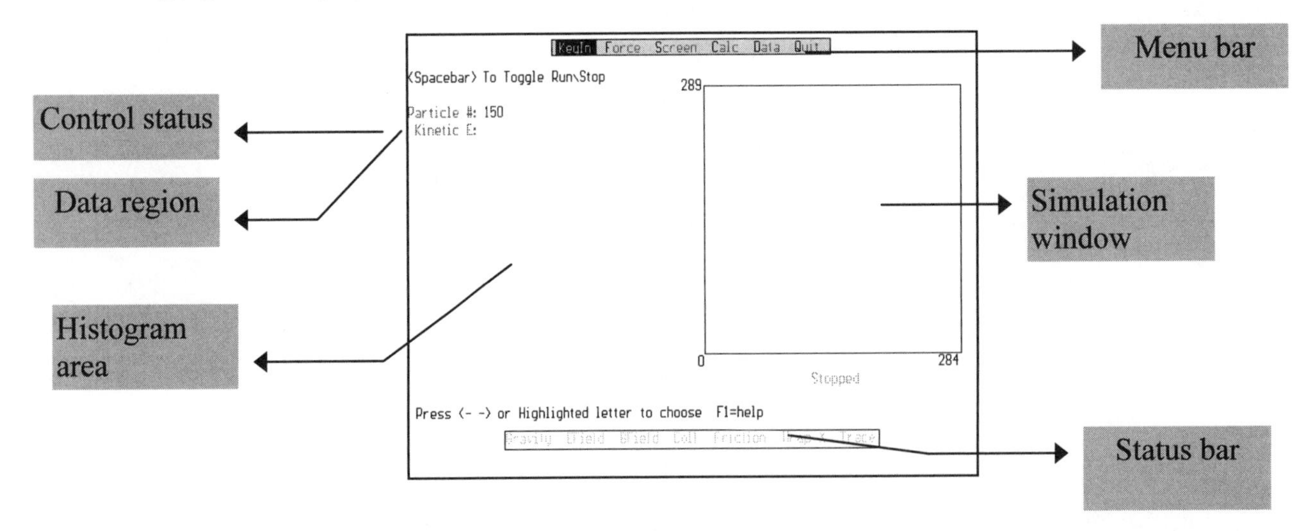

The *control status area* should say "`<Spacebar> to toggle run/stop`". This tells you that to start and stop the simulation, just press `<spacebar>`. If you have not yet done so, do it now.

The program should start a display of 150 particles moving with random velocities. Since the *status bar* is blank, there are no forces on any of the particles.

A. Watch one of the moving particles. Find a slow one so that it is easy to follow. Describe what it is doing.

You can get the program to show you the path followed by one of the particles while making the rest of them invisible. To do this, use the mouse to select `SCREEN` from the menu bar. Press `<enter>` and a menu will pop down.

Choose `TRACE` from the popped down menu either by clicking on it with your mouse or by pressing `<t>` on your keyboard. Press `<enter>` to accept the change of status.

The word `TRACE` should now appear in the status bar and the control status area should say "`<ESC> RETURNS TO SIMULATION`". This is your signal that the computer is still looking for menu entries. Press `<esc>` until the control status area again says "`<Spacebar> to toggle run / stop`". Press `<Spacebar>` to start the program running again.

B. Watch the track of the moving particle. What is it doing?

C. Do you have any evidence of the presence of other particles? Explain.

Return to the menu and by choosing `SCREEN` / `TRACE`, turn *off* the trace feature and return to the full display of all the particles.

D. Consider the distribution of the particles' speeds.

1. Do all the particles move at the same speed in the gas? Explain how you arrived at your answer.

Check your prediction. Turn on the histogram by choosing `SCREEN /HISTOGRAM /ON-OFF` from the menu bar. The program will now display a histogram showing the speeds of the particles. (The title of the histogram says "velocities". **This is an error. Speed ≠Velocity!**)

2. Explain what each bar on the histogram represents.

3. Watch the histogram as the particles move. Does it change? Explain.

4. Was your prediciton in question 1 accurate? Resolve any discrepancies.

III. The Effect of Collisions

The model we've been looking at would be reasonable if molecules were extremely small compared to the separation between them. Then they would almost never run into each other. This is, however, not a reasonable model for the behavior of typical gases that we deal with every day. Molecular separations at Standard Temperature and Pressure (STP) are only about 10 times a molecule's diameter. This means they run into each other fairly often.

We can account for these data by adjusting the parameters of the system. A complete set of new parameters has been saved for you in a file called `settings.txt`. To load this file into the program, choose `Data / Read Settings or Data / Settings` from the menu bar. A dialog box will appear where you can type in the filename where the new settings are saved. We have used the default name for our data file, so you can just press `<enter>` to accept the file followed by `<esc>` until the program returns to run/stop status.

Before starting the program investigate the status of the various system parameters. Note:

- We now have a smaller box (150x150).
- We now have more particles.
- We now have "Collisions" turned on in the status bar.

This last means that two particles in the same pixel on the screen will undergo an energy-conserving collision. The directions of motion and the way the energy is shared between the two particles is changed by a collision.

A. Start the program but stop it again quickly (hit the spacebar quickly!).

1. What is the starting condition for all the particles? Is it different from what you observed on page 73? Explain.

2. What does the speed histogram look like? Is it different from last time? Explain.

B. Start the program again and this time let the proram run.

1. Describe what happens to the distribution of particles in space.

2. Use the histogram to describe what happens to the distribution of particle speed.

3. Is the histogram of speeds similar or different to what you saw before in part IIIA and part II? Give specific descriptions of both similarities and differences.

C. Turn *off* collisions by choosing `FORCE / COLLISIONS` from the menu bar. The status bar should again go blank. Restart the simulation by choosing `SCREEN / RESET / TYPE / MONO CIRCLE` from the menu bar and pressing `<enter>`.

1. Let the simulation run for a few seconds. How is the behavior of the system similar to the one we had before? How is it different?

2. Explain in your own words the reason for the differences in the different cases we have considered.

IV. Mean Free Path

Reconfigure the program by

- turning collisions back on (choose FORCES / COLLISIONS);
- restarting with the expanding circle (SCREEN / RESET / TYPE / MONO CIRCLE).

Let the simulation run until the histogram shows a distribution of speeds and is fairly stable. Then stop the simulation and switch to the mode where the path of a single particle is displayed (use SCREEN / TRACE to do this).

A. Start the simulation and let it run until the particle has hit the wall two or three times. Then stop it by pressing <spacebar>.

1. Describe the motion of your traced particle. Does it move in a straight line? If not, why not?

2. Estimate the average distance your particle traveled between collisions in terms of the units provided.

The average distance that a particle travels between collisions is called the *mean free path*. We can get a calculation of the mean free path in the system by asking the program to keep track of the distance each particle went since its last collision and having it average them.

B. Choose CALC / MEAN FREE PATH from the menu bar and let the program run for a few minutes. How close is the number the program measures from the simulation to your estimate in A.2?

I. Electric Field

Consider a *point charge* with q = +4 units. The "x" indicates the location of a *test charge* with q = +1 units.

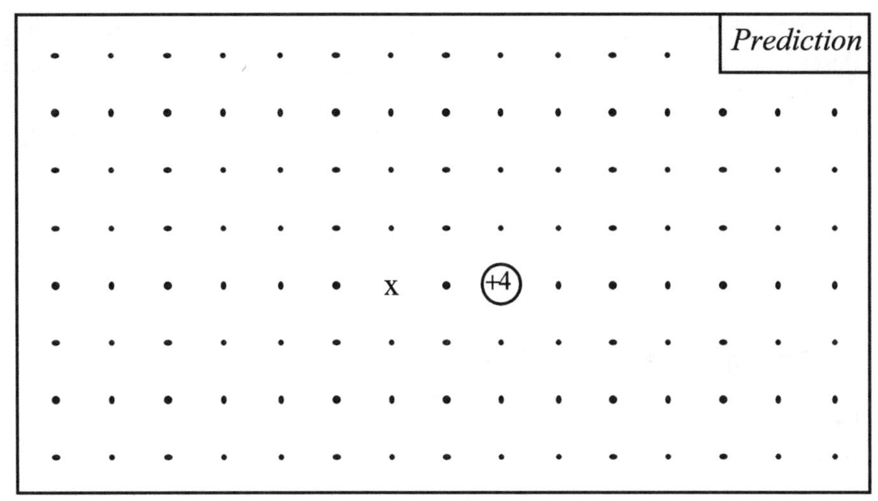

A. In the diagram, draw a vector to indicate the force that the test charge feels when located at point "x." Explain how you arrived at your answer.

1. How, if at all, would your answer change if the *test charge* had charge q = +2 units rather than q = +1 units? Explain

2. How, if at all, would your answer change if the *test charge* had charge q = +3 units rather than q = +1 units? Explain

B. Compare the ratios of the magnitude of the force vector felt by the *test charge* and the strength of the test charge.

The ratio that you have found is called the *electric field* due to a point charge. The electric field describes the ratio of the force felt by a test charge and the strength of the test charge.

C. Does the strength of the electric field depend on the size of the test charge used to measure it? Explain.

II. Strength of the Electric Field Due to a Single Charge

Consider a point charge with q = +4 units.

A. In the diagram, draw vectors to indicate the strength of the electric field at each "x." Explain how you arrived at your answer.

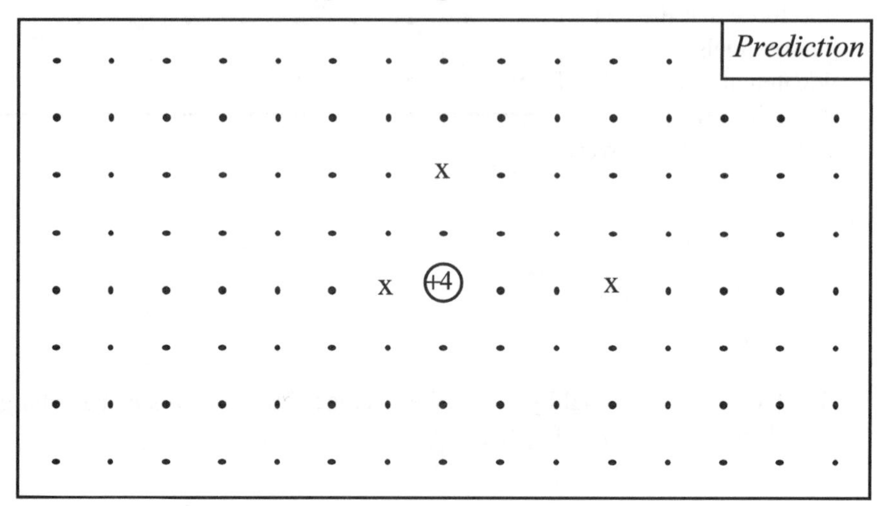

1. How, if at all, would your vectors change if the *point charge* had charge q = +2 units rather than q = +4 units? Explain.

2. How, if at all, would your vectors change if the *point charge* had charge q = –4 units rather than q = +4 units? Explain.

3. How would your vectors change if the *point charge* were moved one grid point to the right? Explain.

Open the Program *EMField.* In the `Sources` menu, choose `3DpointCharges`. If there is no grid on your screen, choose `Display/ShowGrid` from the menu and then `Display/Constrain to Grid`. Drag a single charge of q = +4 units (the solid circles) from the bottom of the screen to the center of the grid.

B. WHILE HOLDING THE LEFT MOUSE BUTTON DOWN AND NOT RELEASING IT, slide the mouse across the screen. Describe what you see on the screen.

1. What does the vector indicate? Explain both magnitude and direction.

2. At each location indicated by the "x" in your prediction on page 78, click the mouse button. Describe what you see on the screen. Resolve any inconsistencies between your prediction and what you see on the screen.

C. How many charges are represented on the screen after you have clicked on each "x"? Explain.

D. Move the point charge one grid point to the right by clicking, dragging, and releasing it at the desired spot. Compare what you see on the screen with your prediction in question A.3. Resolve any discrepancies.

Choose `Sources/3DpointCharges` from the menu. From the charges at the bottom of the screen, drag a charge of q = –4 units to the center of the screen. At the three locations indicated by the "x" in the diagram on page 78, click the mouse button.

E. Compare the vectors on the screen to the vectors described in question B.2. Is your answer consistent with the prediction you made in A.2? Explain.

F. Estimate the length of the three vectors on the screen in terms of grid units. Complete the data table below.

Distance from charge (in grid units)	Length of field vector (in grid units)
1	
2	
3	

Are your values consistent with Coulomb's Law? Resolve any discrepancies.

III. Superposition of Field from Multiple Charges

A. Consider measuring the field from multiple charges. In each of the following cases, predict the magnitude and direction of the electric field. Do not use the program until later questions.

1. Consider a point charge of q = –5 as shown at right. In the diagram, sketch vectors to indicate the electric field at each "x." Explain how you arrived at your answer.

Prediction

x x x

x –5

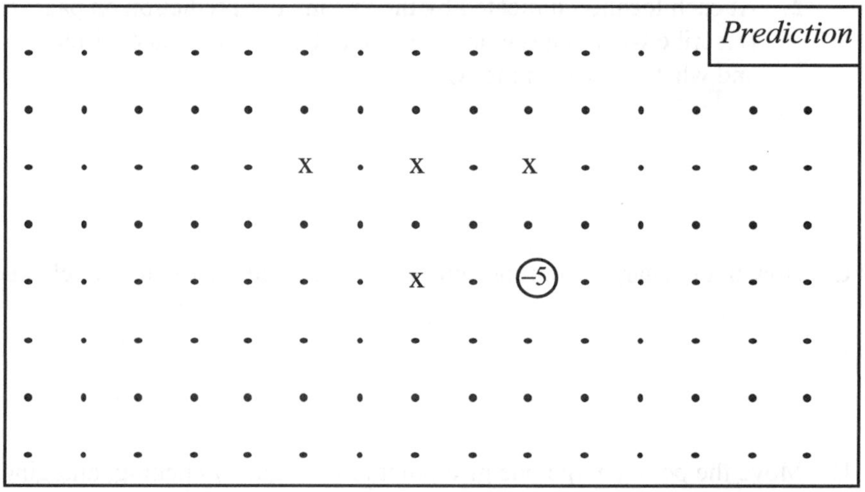

2. Consider a point charge of q = +5 as shown at right. In the diagram, sketch vectors to indicate the electric field at each "x." Explain how you arrived at your answer.

Prediction

x x x

+5 x

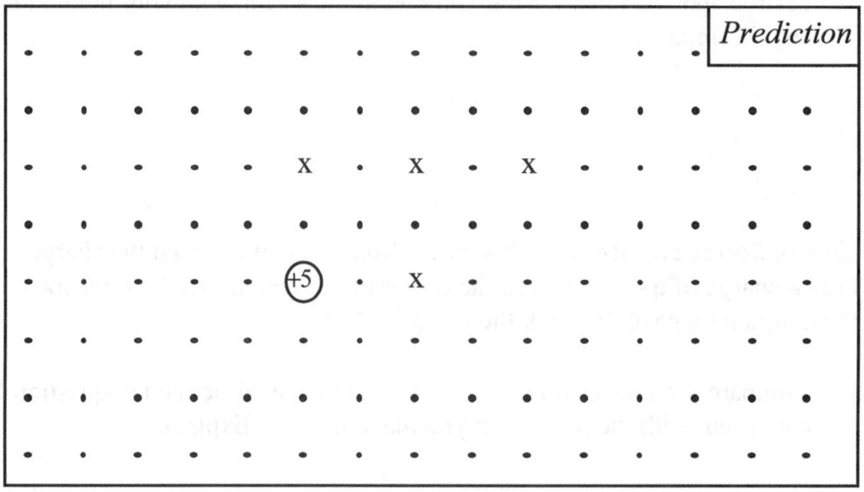

3. In the diagram to the right, sketch the electric field at each "x" when charges of q = +5 units and q = –5 units are on the grid. Explain how you arrived at your answer.

Prediction

x x x

+5 x –5

On your computer, choose `Sources/3DpointCharges` from the menu. Drag charges of q = –5 units and q = +5 units from the bottom of the screen onto the grid and place them as shown above.

B. WHILE HOLDING THE LEFT MOUSE BUTTON DOWN AND NOT RELEASING IT, slide the mouse across the screen. Describe what you see on the screen.

C. At the locations indicated by the "x" in the diagram to the right, click the mouse button. Describe what you see on the screen. Sketch your answer on the diagram.

Measure the length of the top, middle vector on your screen. Also note its direction.

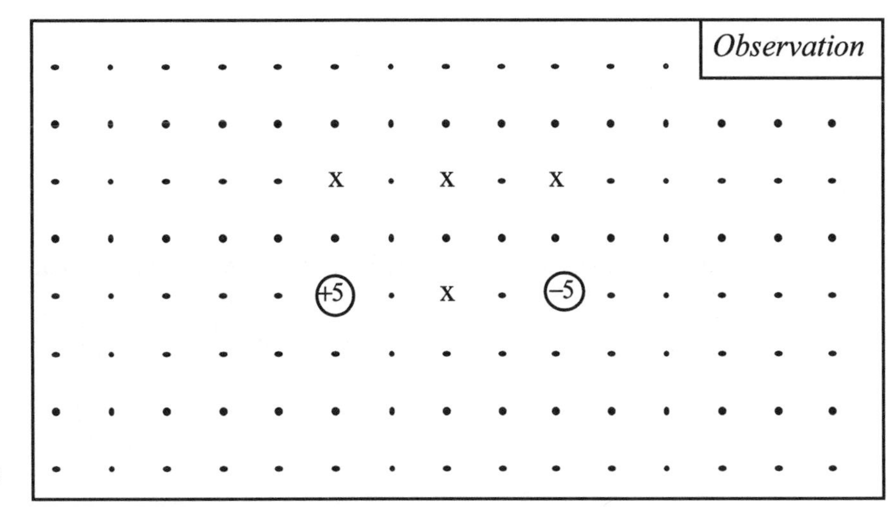

D. Compare the diagram above to your prediction in question A.3. Resolve any discrepancies.

E. Drag the positive charge off the bottom of the screen. Compare the vectors to your prediction in question A.1. Resolve any discrepancies.

Measure the length of the top, middle vector on your screen. Also note its direction.

F. Choose `Sources/AddMoreCharges` and drag a positive charge to the position shown above (the original position of the positive charge). Drag the negative charge off the screen. Compare the vectors to your prediction in A.2. Resolve any discrepancies.

Measure the length of the top, middle vector on your screen. Also note its direction.

G. How are the vectors you have measured in questions E and F related to the field vector for the top, middle "x" which you sketched in question A.3? Include a sketch of the vectors in the space below to explain your answer explicitly.

H. How, if at all, would your responses to questions E, F, and G change if the right point charge had a value of $q = +5$ rather than $q = -5$? Explain how you arrived at your answer.

Use the computer to check your prediction. Resolve any discrepancies.

I. Work and Potential Difference

A charge q_0 is located at position x_A. The charge is able to move along any path to positions x_B, x_C, or x_D.

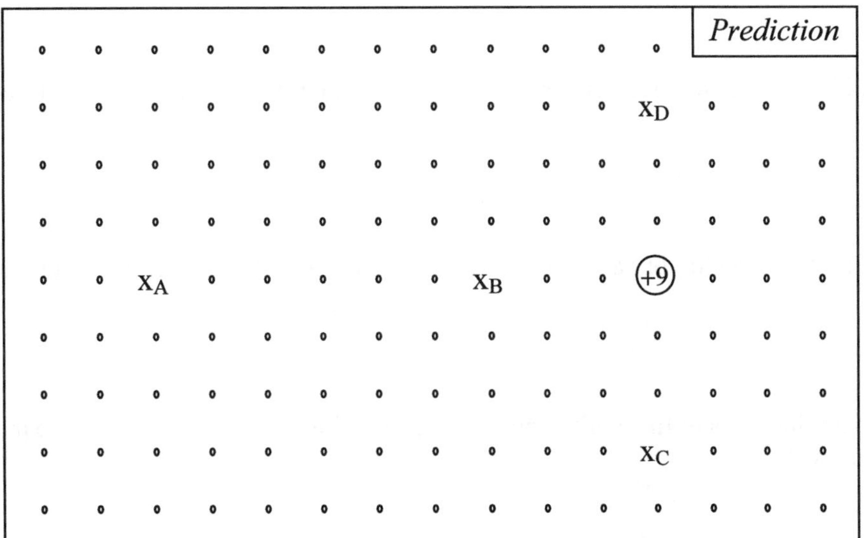

A. Predict which is larger, the work done to move the charge q_0 from rest at x_A to rest at x_B, from rest at x_A to rest at x_C, or from rest at x_A to rest at x_D. Explain how you arrived at your answer. If there is not enough information for you to answer the question, explain what information you would need.

B. Suppose the charge q_0 is moved slowly by an external agent (e.g. a hand) along a straight path from rest at x_A to rest at x_B.

1. Consider the force exerted by the external agent *to move* the charge. (Assume that the acceleration of the charge is near zero at all times.) Is the work done by this force positive, negative, or zero? Explain.

   Compare the sign and magnitude of the work done by the electric force to the sign and magnitude of the work done by the external force. Explain.

2. Write an expression that describes the work done by the electric field of the point charge on the charge q_0 as it moves from rest at x_A to rest at x_B. Explain.

3. Write an expression that describes the work done by the external agent on the charge q_0 as it moves from rest at x_A to rest at x_B. How did you arrive at your answer?

4. How, if at all, would your answer to question 3 change if the charge had magnitude $2q_0$?

   How, if at all, would your answer to question 3 change if the charge had magnitude $3q_0$?

5. Compare the ratios of the work done by the external force on the charge to the magnitude of the charge.

C. The ratio you have found is the *electric potential difference,* ΔV. The electric potential difference between two points describes the ratio between the work done on a test charge to move it from rest at one point to rest at the second point and the magnitude of the test charge.

1. Does the potential difference depend on the magnitude of the test charge? Explain.

2. Consider that the charge is moved from rest at x_A to rest at x_B along the shortest possible path. Compare the sign and magnitude of the potential difference for this path to the sign and magnitude of the potential difference when the charge is moved from x_B to x_A. Explain.

3. Find a value (in terms of k_e) for ΔV between locations x_A (9 grid points from the q = +9 charge) and x_B (3 grid points from the q = +9 charge). Assume the units of charge and distance are Coulomb and meter, respectively. Show all work.

II. Potential Difference of a Point Charge

Start the program *EMField* from the desktop. Select `Sources/3DPointCharges` from the program's menu. Choose `Display/ShowGrid` from the program's menu, and then choose `Display/ConstrainToGrid`. Place a single positive charge of magnitude +9 at a grid point near the right side of the screen (about 2/3 of the way over) and about half way down. Select `FieldAndPotential/PotentialDifference` from the menu.

A. WHILE HOLDING THE LEFT MOUSE BUTTON DOWN AND NOT RELEASING IT, slide the mouse from the left edge of the screen toward the charge of q = +9 units. RELEASE the mouse at a point close to but not on top of the charge. Describe what you see on the screen.

1. Choose `Display/CleanUpScreen`. On the computer, move the mouse from x_A to x_B along a straight path by clicking and holding your mouse button at x_A and releasing it at x_B.

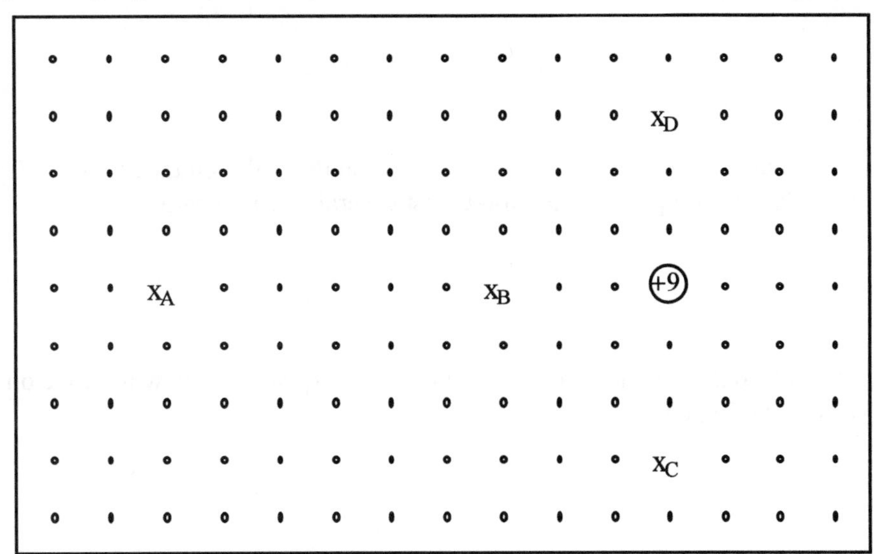

2. Compare the value for the potential difference displayed on the computer with the value you computed on p. 84. Resolve any discrepancies.

B. While clicking the mouse and holding the mouse button, move from x_B to x_A and release the mouse button at x_A.

1. How does the value for ΔV between x_B and x_A compare to the value between x_A and x_B? Explain.

2. How is the sign of ΔV reflected in the color of the line drawn on the screen?

3. Does the value of the potential difference between x_B and x_A depend on the path along which you move? Explain how you arrived at your answer.

C. Is the potential difference between positions x_A and x_C *greater than, less than, or equal to* the potential difference between positions x_A and x_B? Explain.

1. Choose `Display/CleanUpScreen`. Use the mouse to find the potential difference between positions x_A and x_C. Compare the value of ΔV between x_A and x_C to the value of ΔV between x_A and x_B.

2. Does the value of ΔV between x_A and x_C depend on the path you take to move from x_A to x_C? Use the computer simulation to test several possible paths.

3. How, if at all, could you move a particle with charge q_0 so that the work done on the particle is always zero? Explain.

4. How, if at all, could you draw a path between points x_B and x_C along which the potential difference is always zero? Explain.

5. Choose `Display/CleanUpScreen`. Use the mouse to find a path between x_B and x_C along which ΔV is always zero. Compare this path to your prediction. Resolve any discrepancies.

D. What is the potential difference between points x_A and x_D? Use the computer simulation to check your prediction. Resolve any discrepancies.

A. Two small metal spheres are hung from the ceiling by nylon (insulating) thread, so that they hang straight down and just barely touch each other. A charged rod is briefly brought into contact with one of the spheres. After the rod is removed, the two spheres move away from each other, until they eventually come to rest while hanging 8 degrees from the vertical. What is the charge on each sphere? You know that the nylon threads are 1 meter long, and you measured the mass of the spheres on a scale beforehand and know that each is 100 g.

B. Two charges are placed a distance *d* apart. One is a +8 μC charge, and the other is a +2 μC charge. Both are fixed so they can't move. Where could you place a third charge (of +1 μC) so that it stays where you place it?

How, if at all, would your answer change if the +2 μC charge were changed to a –2 μC charge?

A. Consider the circuit shown at right. Assume that the wires battery are ideal. It is observed that when the switch is closed, the bulb is initially not lit, and then gradually increases in brightness. Once the brightness of the bulb is the same as its brightness when connected directly to the battery, it remains unchanged.

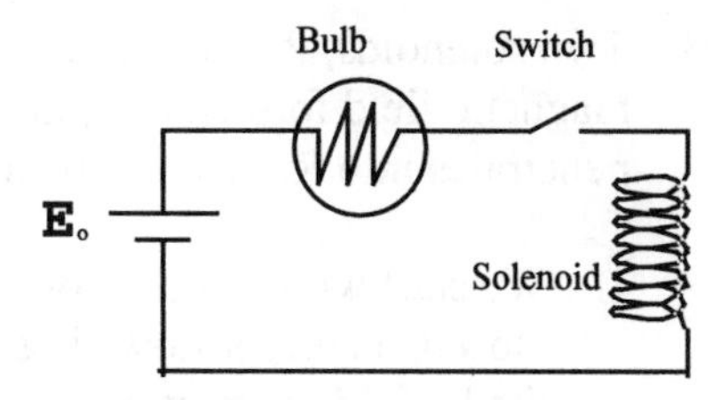

1. Just after the switch is closed, what is the voltage across the bulb? Explain how you know.

2. Just after the switch is closed, what is the voltage across the solenoid? Explain how you know.

3. After a long time, what is the voltage across the bulb? Explain how you know.

4. After a long time, what is the voltage across the solenoid? Explain how you know.

B. Two solenoids, P and R, are sufficiently close together that the magnetic field formed in P, in the presence of the electric current, also penetrates into R. See figure at right.

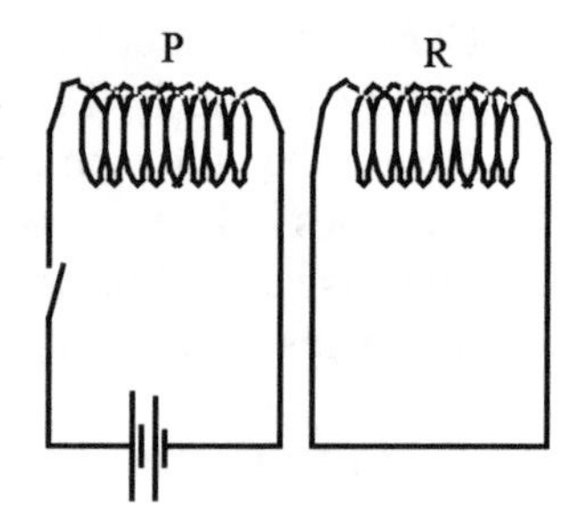

1. We start with switch S open. At the instant that the switch is closed, a current flows in solenoid P. Establish the direction of the B-field penetrating into solenoid R and show its direction. Explain how you arrived at your result. If the B-field is zero, say so explicitly and explain your reasoning.

2. Just after the switch is closed, what is the direction of the induced current in R? Explain how you arrived at your result. If the current is zero, say so explicitly and explain your reasoning.

3. After a long time has passed, there is a steady-state current in solenoid P. Describe what, if anything, happens in R, showing the direction of any possibly induced current. Explain how you arrived at your conclusion. What is the situation in R after the current in P has dropped to zero? Explain your reasoning.

4. Suppose the battery in the diagram above is replaced with an AC voltage source such that the current, I_P, through solenoid P is shown at right. Sketch the current in solenoid R as a function of time on this graph and explain your reasoning. Note that a positive current on the graph indicates a clockwise current around a circuit

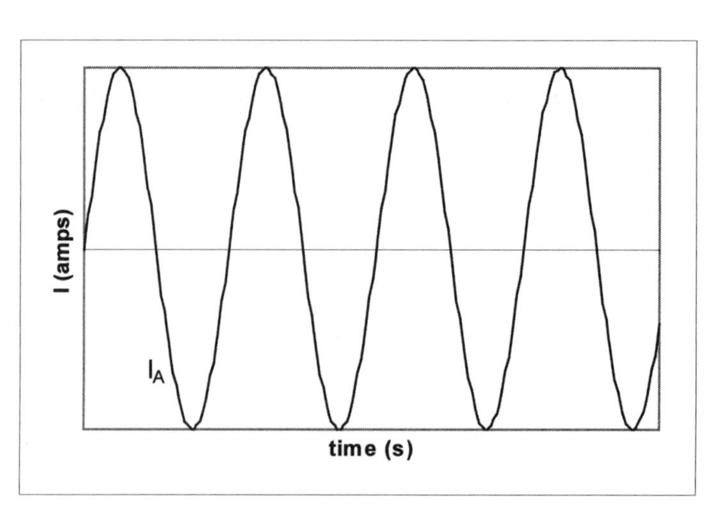

C. Consider the circuit shown at right. Assume that the resistance across the solenoid and the wires is zero and there is an initial charge, Q, on the capacitor.

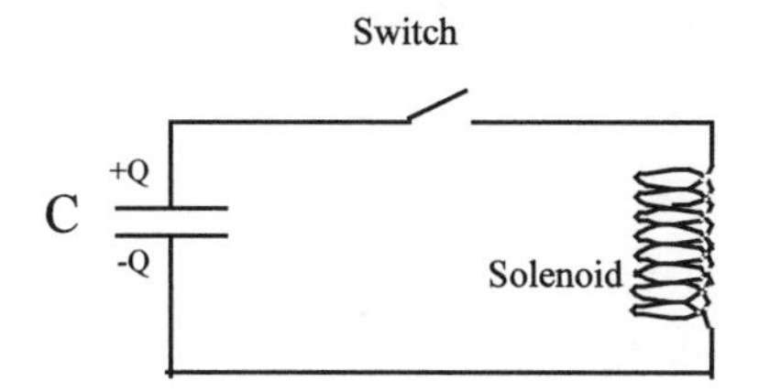

1. At the instant after the switch is closed:

   a. What is the voltage across the capacitor and the voltage across the solenoid? Explain how you know.

   b. Describe the current in the circuit. (In which direction, if any, is it? Is it increasing or decreasing?) Explain using Lenz's Law.

2. At some later time, t_1, the charge on the capacitor has been reduced to Q/2.

   a. What is the voltage across the capacitor and the voltage across the solenoid? Explain how you know.

   b. Describe the current in the circuit. Explain using Lenz's Law. (Hint: Does the EMF across the solenoid correspond to an increasing current or a decreasing current?)

3. At some later time, t_2, the charge on the capacitor is 0.

   a. What is the voltage across the capacitor and the voltage across the solenoid? Explain how you know.

   b. Describe the current in the circuit using Lenz's Law. Explain.

4. Describe what happens in the circuit after t_2. Explain your reasoning.

Tutorial Homework

Two cars starting from rest at the same position are facing the same direction on a long, straight road. The first car starts moving at t = 0 seconds and accelerates at 5 m/s^2 for 2 seconds then travels at a constant speed. The second car starts moving 3 seconds after the first car started and accelerates at 20 m/s^2 for 1 second and then travels at constant speed.

A. In the space below sketch a position vs. time graph for each car, labeling each car's curve. Explain how you determined the shape of the graph.

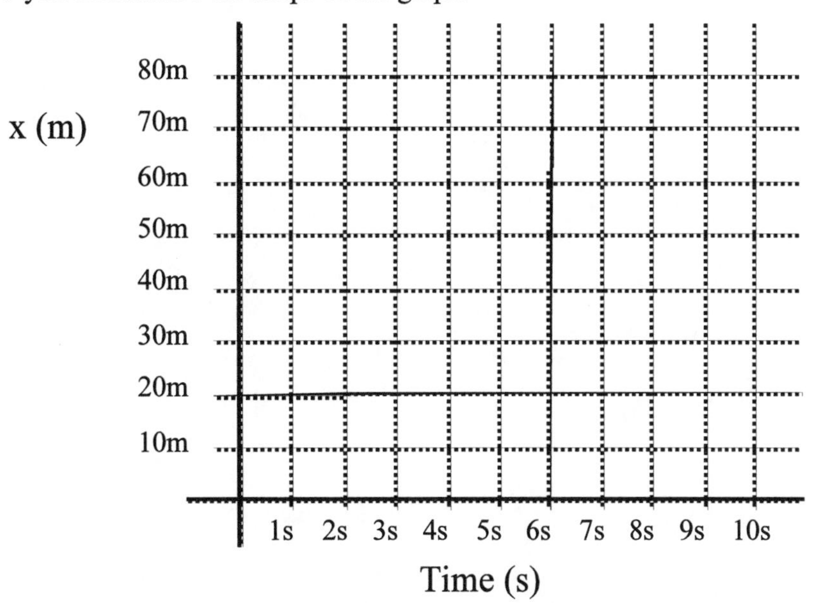

B. Does the second car pass the first car in the first 10 seconds? If the second car does pass the first car, state when and where this occurs, and explain how you know. If the second car does not catch up explain how you know.

C. Are the two cars ever at the same speed at the same time after t = 0 seconds? Explain how you could tell from the graph.

A. Consider a block of wood given an initial push and then sliding up an incline. It reaches the top of the incline, turns around, and slides down the incline. There is friction between the block and the incline.

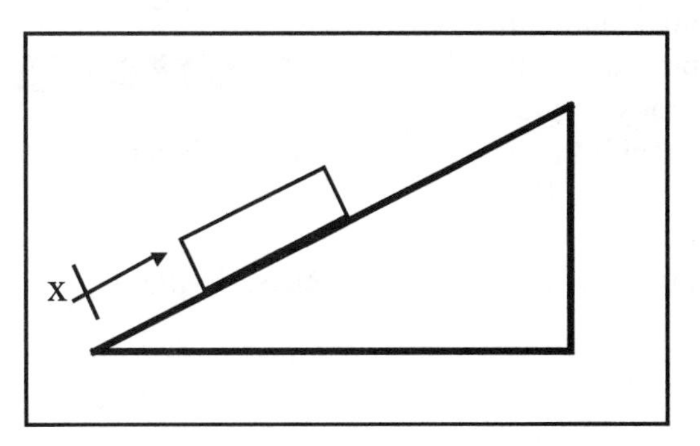

There is friction.

1. In the space below, sketch a free-body diagram for the block

   a. on its way up the incline and

   b. on its way down the incline.

Label all forces clearly.

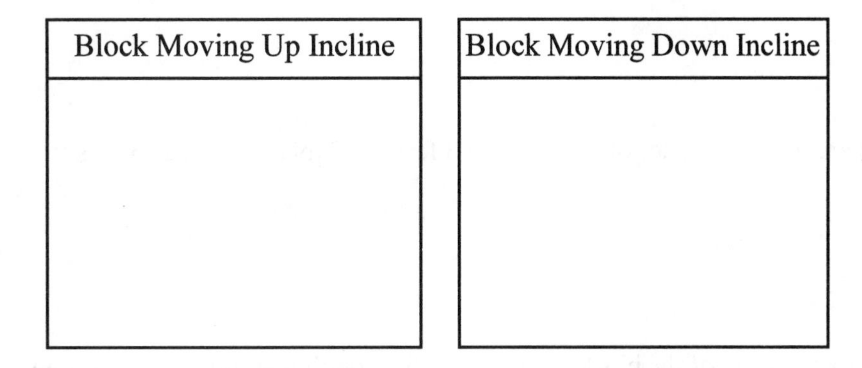

2. In the space below, sketch the velocity vs. time and acceleration vs. time graphs for the block of wood during its *ENTIRE* motion up and down the incline.

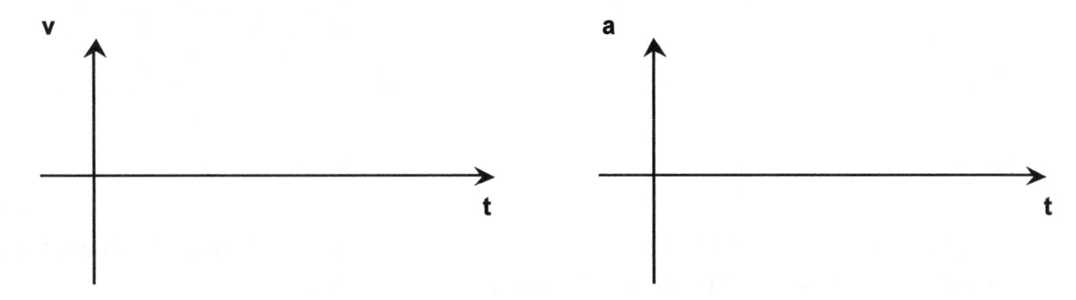

Explain how you determined the shapes of your graphs.

B. You are pushing horizontally on a block that is resting on a table.

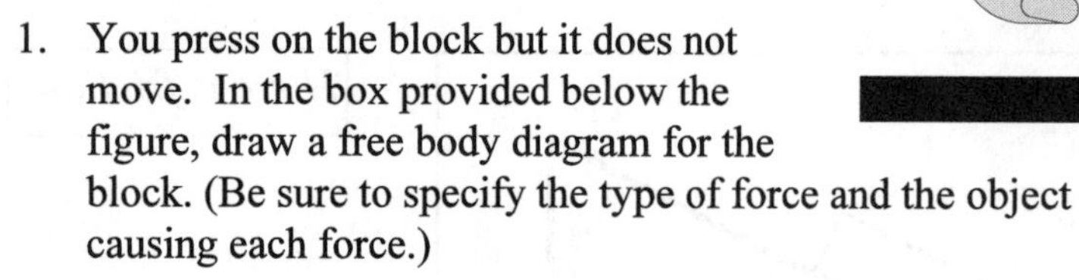

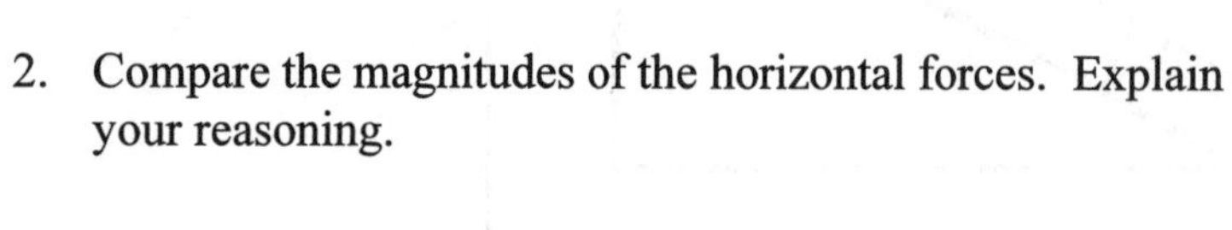

1. You press on the block but it does not move. In the box provided below the figure, draw a free body diagram for the block. (Be sure to specify the type of force and the object causing each force.)

2. Compare the magnitudes of the horizontal forces. Explain your reasoning.

3. You press a bit harder and then the block moves with a constant velocity. Draw a free body diagram below for the block when it has a constant velocity.

4. Compare the magnitudes of the horizontal forces. Explain your reasoning.

5. Suppose the mass of the block is 0.4 kg and the coefficient of friction between the block and the table is 0.3. What force will you have to use to keep it going at a constant velocity of 0.2 m/s? (You may take g = 10 N/kg.)

6. If you now push with twice the force as in the previous question, how if at all does the motion of the block change? Explain your reasoning.

A. Two boxes of unequal mass ($m_A < m_B$) are held by a hand as shown. The hand is moved upward at an *increasing speed.*

B
A

1. In the space below, sketch a free-body diagram for Box A and a separate free-body diagram for Box B as the hand is speeding up. Label all forces clearly.

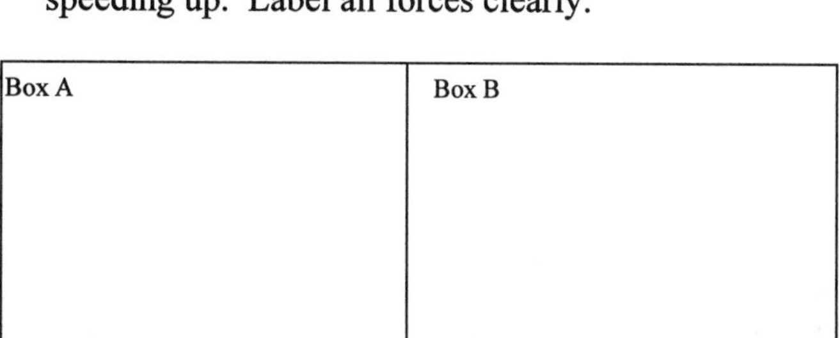

2. Consider the following **incorrect** analysis of the situation above by a student:

   "Box B is pushing down on box A because of its weight. So the weight of box B acts on box A."

   a. Why is the analysis incorrect? Explain.

   b. Under what circumstances is the force exerted by box B on box A equal in magnitude to the weight of box B? Under what circumstances are the magnitudes of those forces unequal?

B. Two gliders of unequal mass ($m_A < m_B$) glide toward each other on a frictionless air track so that they collide.

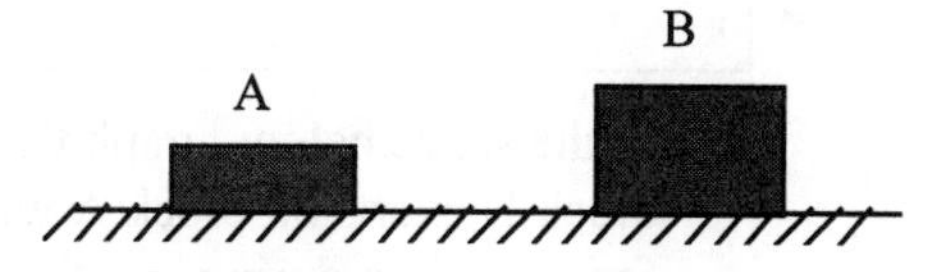

1. In the space below, sketch a free-body diagram for Glider A and a separate free-body diagram for Glider B while they are colliding.

Glider A	Glider B

a. Compare the horizontal forces in your free body diagrams. Explain your reasoning.

b. Compare the acceleration of Glider A during the collision to the acceleration of Glider B. Explain your reasoning.

C. Two blocks of unequal mass ($m_A<m_B$) are placed in contact on a surface. Block A is pushed horizontally by a hand. The situation is shown at right, along with a graph of the acceleration vs. time for the two blocks.

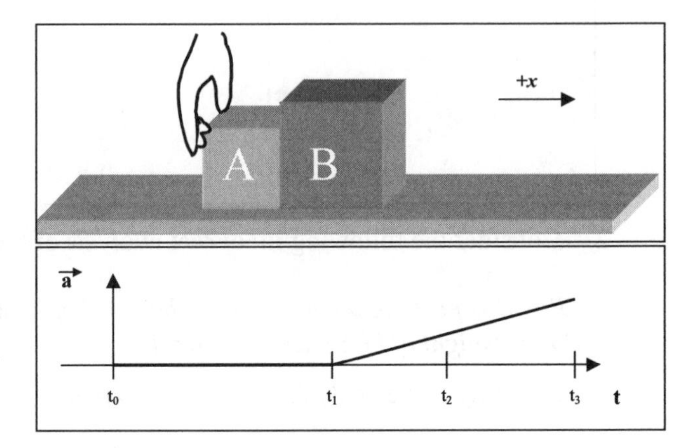

1. In the space below sketch a free body diagram for each block at $t = t_2$. In each diagram describe each force and label each force as you have done in the tutorial.

Block A	Block B

2. In the space below, graph the force block A exerts on block B with a solid line and the force block B exerts on block A with a dotted line, for the motion from t_0 to t_3. If either of the forces is zero during any time interval, state so explicitly. Explain your reasoning.

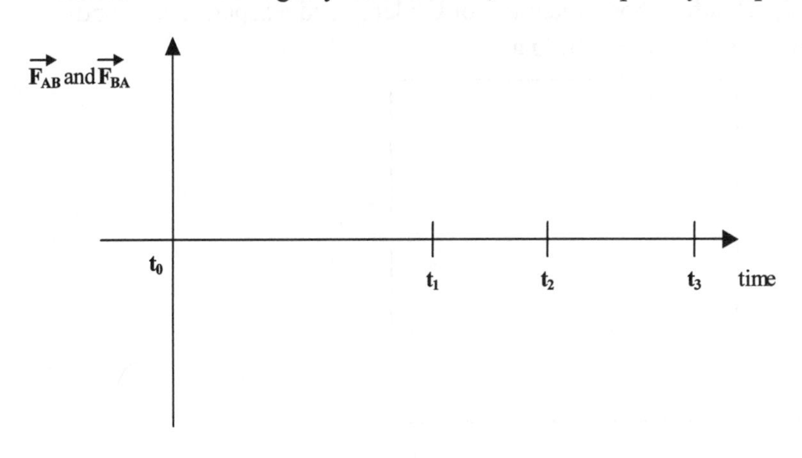

A. A block is pulled along a horizontal surface by a constant force **T** as shown in the diagram. From point A to point B (a distance of 1 m) there is no friction; from point B to point C (also 1 m distance) there is kinetic friction with coefficient μ. The block starts from rest at point A and comes again to rest at point C. The force exerted by the tension force is exerted throughout the motion of the block.

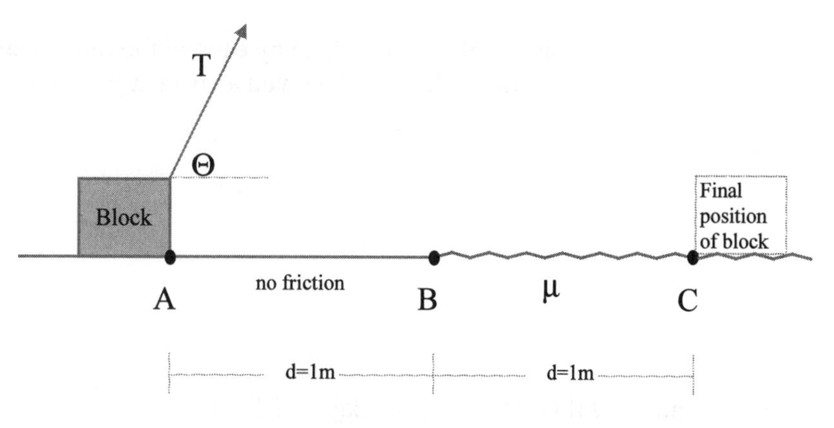

1. Draw a free-body diagram for the block when it is at a point between A and B. Label all forces clearly.

2. Draw a free-body diagram for the block when it is at a point between B and C. Label all forces clearly.

3. Is the magnitude of the net force acting on the block when it is between A and B *greater than*, *less than*, or *equal to* the magnitude of the net force acting on the block when it is between B to C? Explain your reasoning.

4. Rank the magnitudes of the work done by each of the forces acting on the block as it moves from point B to point C. Explain how you arrived at your answer.

5. Calculate μ if Θ=60°, m_{block}=1.5kg, and T=5N.

A. Consider the motion of a projectile in two dimensions. Suppose you are standing on level ground and are throwing a ball at an angle of about 45°. **For this problem, ignore air resistance.**

1. On the diagram below, draw a picture of the path the ball would follow.

2. On the picture draw images of the ball:

   - part way between the start and the highest point
   - at the highest point
   - part way down between the highest point and when it hits the ground.

   On each of these images, draw arrows to indicate the velocity of the ball at the instant when the ball is at the position shown by the image. Use single line arrows (→).

3. On the figure below, copy your trajectory and ball images. On each of the ball images draw a free-body diagram to show all the forces on the ball at the instant when the ball is at the position shown by the image. Label all forces by identifying: a) the type of force, b) the object on which the force is exerted, and c) the object exerting the force.

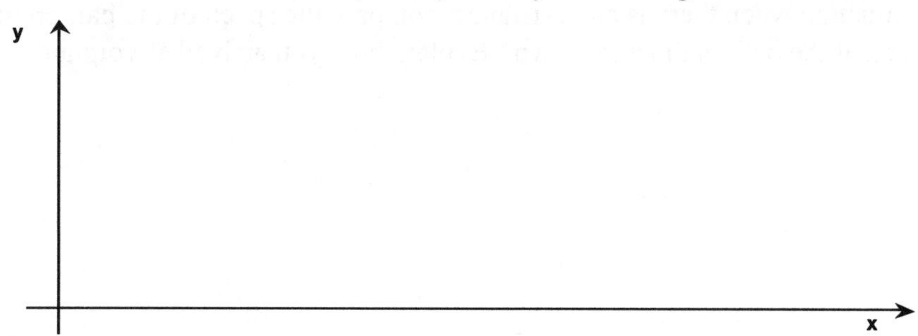

B. Now suppose **there is air resistance.**

1. On the figure below, copy your trajectory and ball images assuming that the trajectory *would be the same as if there were no air resistance* (this is not really the case).

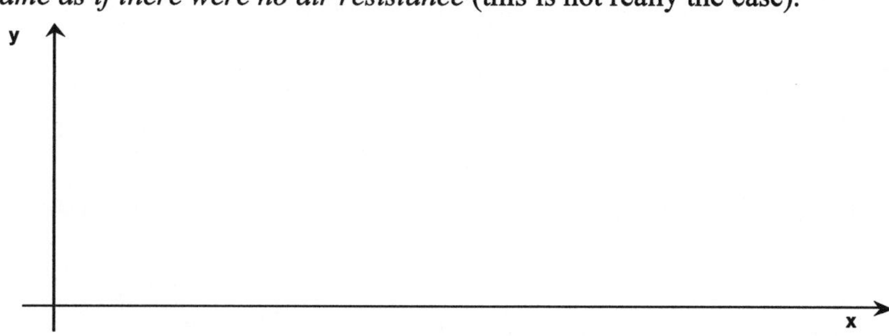

On each of the ball images on the picture on the previous page, draw free-body diagrams to show all the forces now acting on the ball at the instant when the ball is at the position shown. Label all forces by identifying: a) the type of force, b) the object on which the force is exerted, and c) the object exerting the force.

2. From the way the forces look at the three points, argue what the effect of air resistance will be on each of the following. For each answer, **explain your reasoning**.

   a. the height to which the ball rises

   b. the total distance the ball travels

3. In the situation when there is air resistance, compare the speed of the ball on the way up vs. the speed of the ball on the way down. Explain how you arrived at your answer.

A. A cylinder is hung from a spring which is attached to a frame (see figure). The cylinder is pulled downward a distance y_{pull} and released. At the instant the cylinder passes its equilibrium position (as defined in the tutorial), a clock is started ($t = 0$).

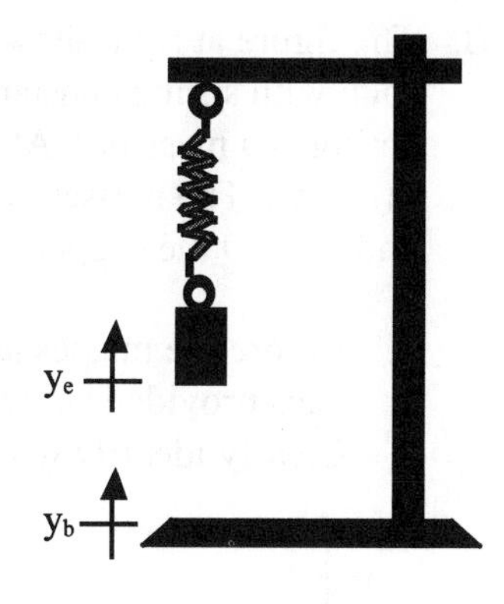

Consider ***two*** coordinate systems to describe the motion of the cylinder. The first coordinate system is chosen with an origin ($y_b = 0$) at the base of the frame, and the upward direction is considered positive. The cylinder is shown at rest at its equilibrium position, y_0. The second coordinate system measures displacement from the cylinder's equilibrium position ($y_e = 0$).

1. On the axes below, sketch graphs of y_b vs. t and y_e vs. t.

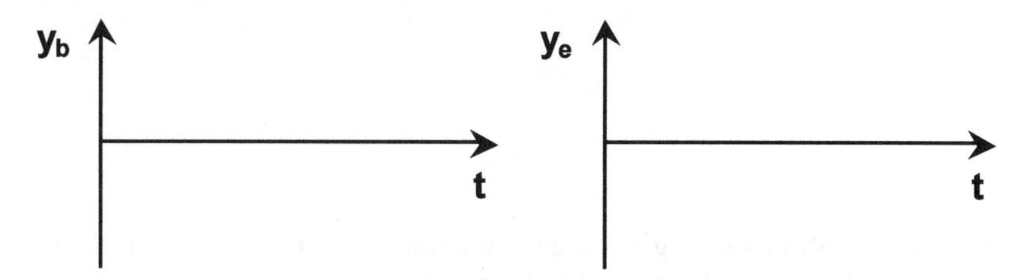

Account for any differences between the two graphs.

2. Write the general equation that gives y_e as a function of time for the y_e vs. t graph you sketched above.

3. Write the equation that gives y_b as a function of time. Explain how you arrived at your answer.

4. In the box at right, sketch a free-body diagram for the instant in time when the cylinder is located at $y_e = 0$. Label all forces like in tutorial. Are the forces the same in both coordinate systems?

5. Use your equations above to show that Newton's Second Law is the same in both coordinate systems. Show all work.

B. The figure at right shows two identical, massless, frictionless, springs, each with spring constant k, hanging from a bar. Attached to one spring is a mass m_1. Attached to the other spring is a mass m_2, where $m_2 > m_1$. The masses are supported so that the springs are unstretched, and at $t = 0$ the support is removed so that the masses begin to fall.

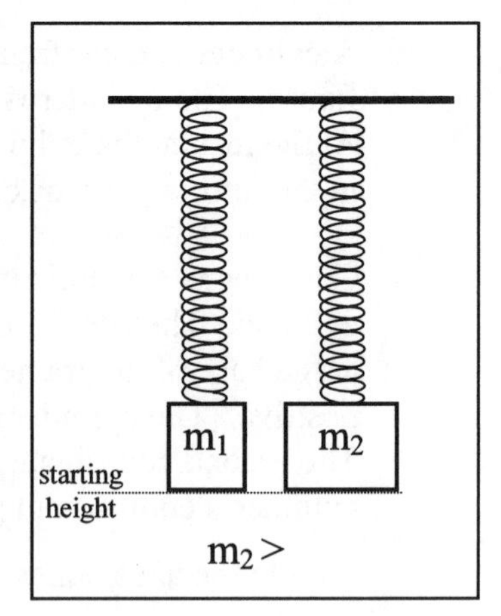

1. Before the masses are released, they are at height $y = 0$. On the axes provided below, graph the height of each of the masses. Clearly identify which curve corresponds to which mass.

2. Determine the equation which gives y as a function of t for m_1. Explain how you arrived at your answer. (Hint: To find the amplitude, consider the location where the net force is zero.)

3. Determine the equation which gives y as a function of t for m_2. Explain how you arrived at your answer.

4. If $m_1 = 50\text{g}$, $m_2 = 200\text{g}$, and $k = 50$ N/m, find the difference in height between the two cylinders at time $t = 6.28$ sec. Show all work.

A. A method for generating a wavepulse is to move one end of a spring quickly up a distance *d* and then back down (see figure). The hand takes the same amount of time to move up as to move down. Consider a second wavepulse generated with the same amplitude, *d*, on a different spring (spring 2). It is observed that the wave speed on spring 2 is half that in the original spring (spring 1).

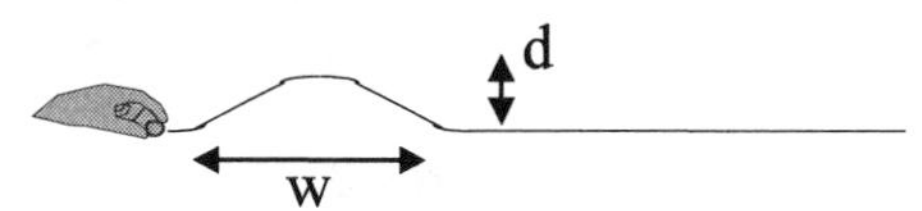

1. How can you account for the difference in speed of the wavepulse on the two springs? Explain.

2. What could you change about the creation of the second wavepulse or spring 2 so that the wavepulse on spring 2 traveled at the same speed as the wavepulse on spring 1? Explain.

B. The pulse shown in the figure at the right is moving to the right at 50 cm/s.

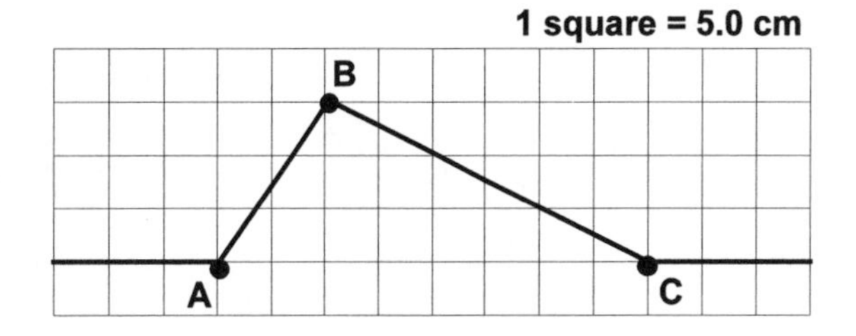

1. Draw velocity vectors to indicate the instantaneous velocity of the piece of spring located at the horizontal midpoint of each square.

2. Using qualitative reasoning explain how the velocity of a piece of spring is related to the slope of that piece of spring.

C. The diagram below shows two wavepulses moving toward each other on the same side of a spring at time t = 0 sec. Each pulse is moving at a speed 100 cm/sec. Each block represents 1 cm.

1. In the grids provided to the right, sketch a sequence of diagrams that show both the positions of the individual pulses (with dashed lines) and the shape of the spring (with a solid line) at 0.02 sec intervals.

2. Draw velocity vectors to indicate the *velocity* of the piece of spring located at the horizontal midpoint of each square at time t = 0.04 sec. Explain how you arrived at your answer.

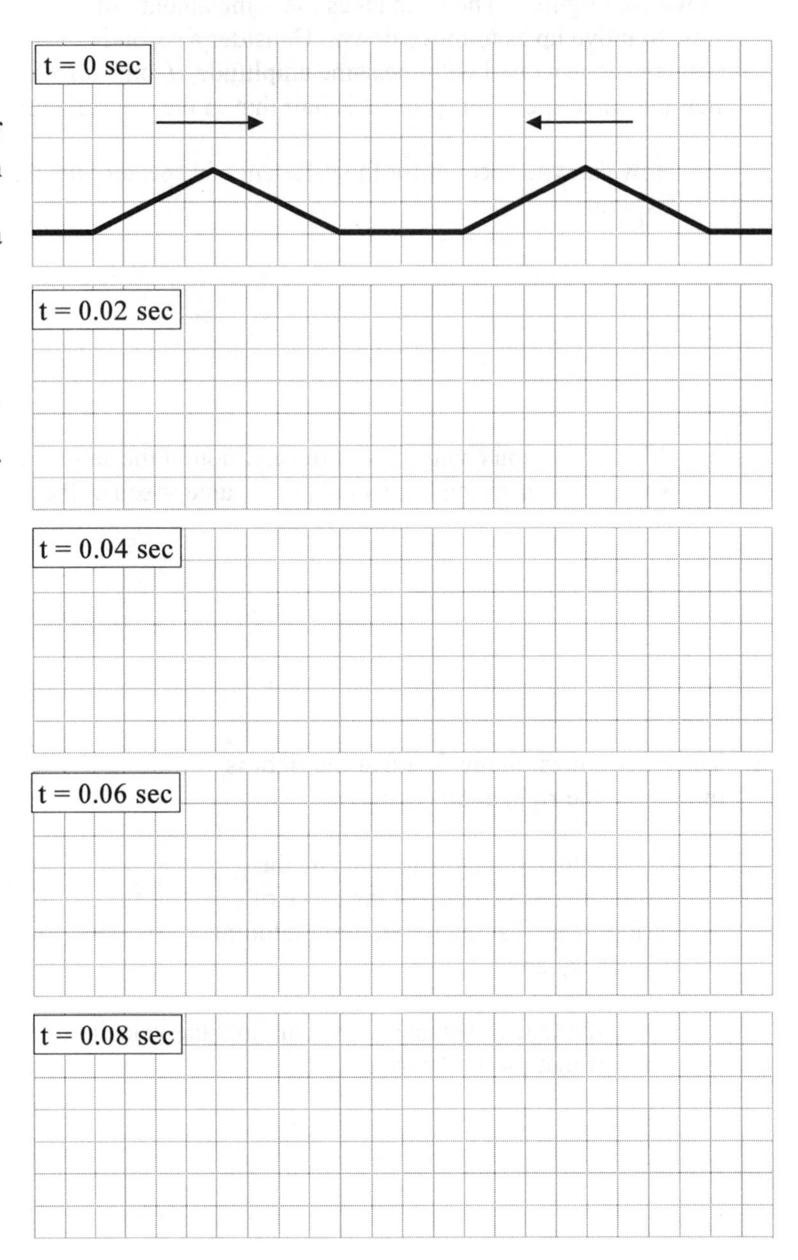

D. Two infinite (continuing in both directions) waves are traveling along a taut spring of uniform mass density. At time t = 0 seconds, the waves have the same shape and are in the same location. One is traveling to the *right*, the other is traveling to the *left*. *One* of the waves is shown in the space below. (At time t = 0 sec, the other wave's peaks perfectly overlap the first wave's peaks.) In the diagram, each block represents 10 cm. After t_0 seconds, the wave traveling to the *right* has traveled 20 cm.

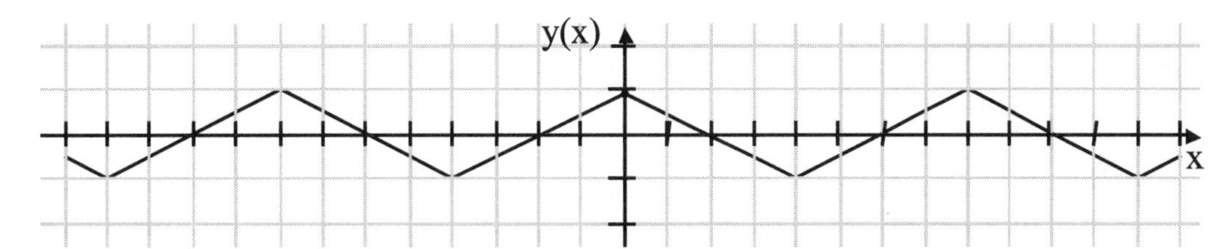

1. Compare the speed of the two waves. Explain how you arrived at your answer.

2. Using two different colors of pen or pencil, sketch each individual wave at time t_0 in the graph below. (Do not sketch the shape of the spring, just each wave.) Indicate the direction that each wave is traveling on your sketch.

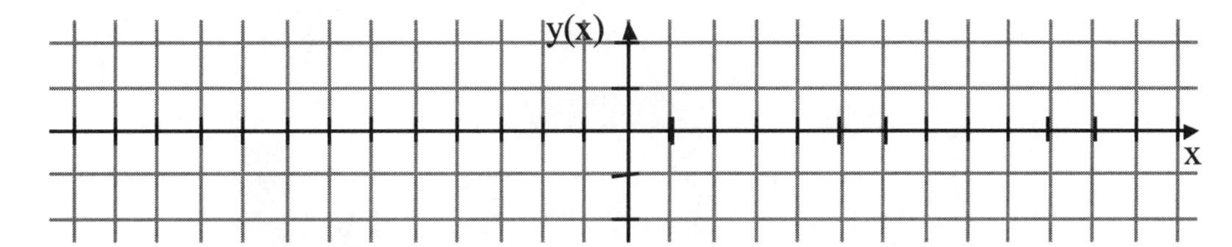

3. In the graph below, sketch the shape of the spring (with both waves traveling on it) at time t_0. Explain how you arrived at your answer.

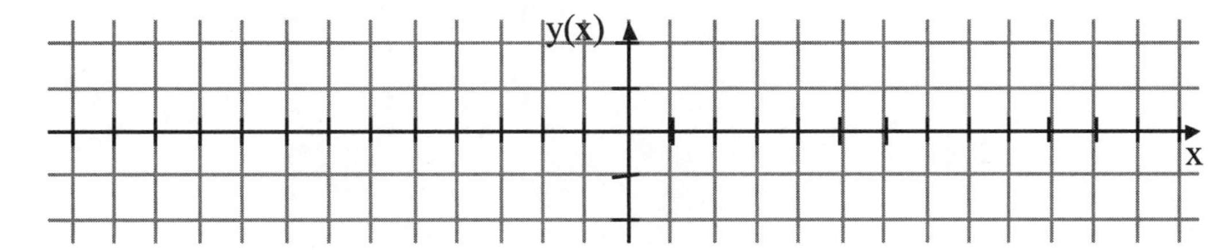

4. If t_0 = 0.2 sec, find the velocity of the piece of spring located at x = 0 at the instant you have drawn in question 3. Explain how you arrived at your answer.

A. Suppose a transverse pulse is propagating along a spring resting on a very smooth surface. The velocity of the pulses is 600 cm/sec *to the left*. At time t = 0 sec, the displacement of the spring from its equilibrium position can be written as a function of x, $y(x) = \frac{50cm}{(x/b)^2 + 1}$, where b = 20 cm.

1. Write the equation that describes the displacement of the spring from its equilibrium position at any position, x, and at any time, t. Explain how you arrived at this answer.

2. Compare your equation to the equation which you derived in section I.B, question 9 of the tutorial. What has changed? Explain the effect of this change.

3. In the space below, graph the displacement of the spring after 0.1 seconds have passed.

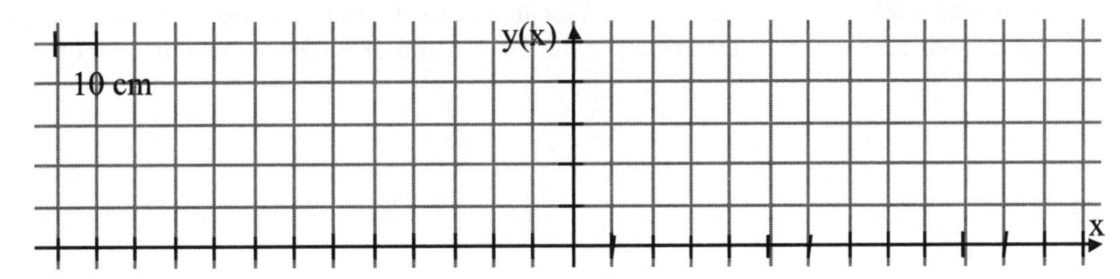

4. At $t = 0.1$ sec, what is the displacement at a point 20 cm to the left of the location of the peak? Explain how you arrived at your answer.

5. What point(s) had this same displacement at time t = 0 seconds? Explain how you arrived at your answer.

B. Consider a sinusoidal wave traveling *to the right* with a speed of 600 cm/sec. The equation describing the displacement of the spring from equilibrium at time t = 0 sec is $y_2(x) = A\sin(\frac{2\pi}{\lambda}x)$. The wave has a wavelength, $\lambda = 80$ cm, and an amplitude, $A = 20$ cm.

1. Sketch the shape of the spring at time t = 0 in the graph below. Use the indicated scale.

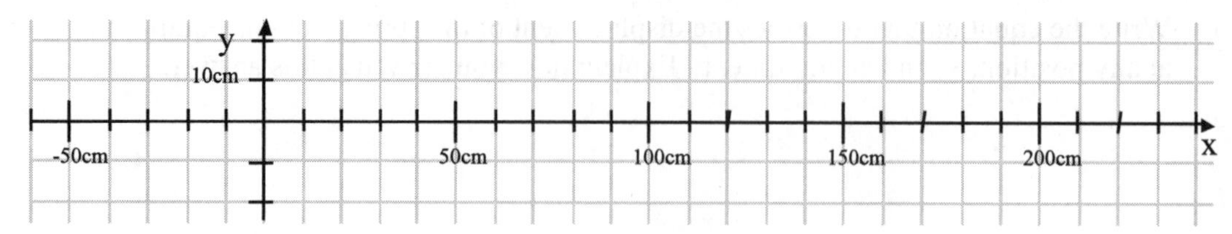

2. Write an equation that describes the displacement of any piece of the spring at any time for this pulse. Explain how you arrived at this answer.

3. Compare the motion of the piece of spring located at x = 0 cm to the motion of the piece of spring located at x = 40 cm. Describe the similarities and/or differences of the two cases.

4. Write the equation that describes the motion of the piece of spring located at x = 40 cm as a function of time. Explain how you arrived at your answer.

C. Consider a transverse pulse traveling *to the right* with a speed of 600 cm/sec. The equation describing the displacement of the spring from equilibrium at time t = 0 sec is $y_1(x) = A_1 e^{-\left(\frac{x}{b_1}\right)^2}$, where A_1 = 20 cm and b_1 = 20 cm.

1. Sketch the shape of the spring at time t = 0 sec in the graph below. Use the indicated scale.

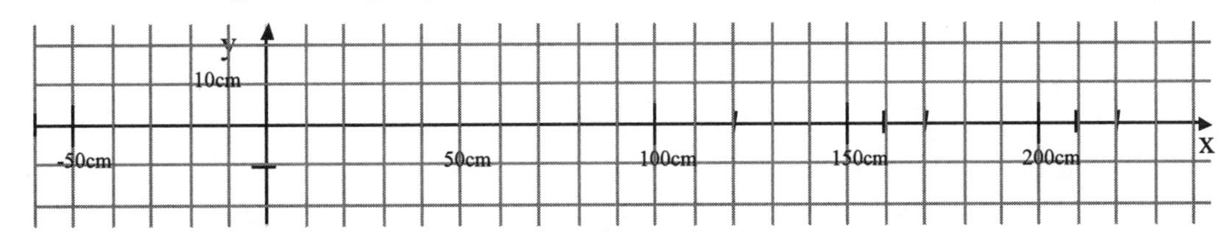

2. Write the equation that describes the displacement of any piece of the spring at any time for this pulse. Explain how you arrived at this answer.

3. Now suppose a second transverse pulse moving *to the left* is also present on the spring. At t = 0 sec the equation describing this 2nd pulse is $y_2(x) = A_2 e^{-\left(\frac{x-200\,cm}{b_2}\right)^2}$ where A_2 = 10 cm and b_2 = 10 cm. Sketch the shape of the spring (with both wavepulses!) at time t = 0 sec in the graph below, labeling the pulse going to the right "1" and the pulse going to the left "2".

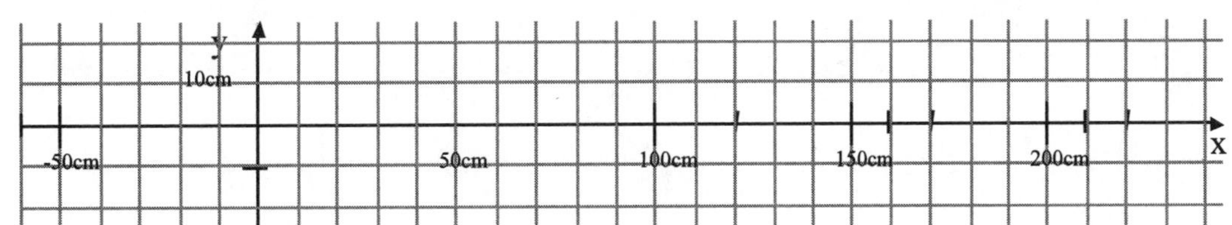

4. At what time and position do the maxima of the two pulses meet? Show your work.

5. Write the equation that describes the shape of the spring when the maxima meet. Show your work.

A. Consider a single dust particle floating a distance x_0 from a loudspeaker (see figure). The loudspeaker is turned on and produces a sound with a constant frequency f. The speed of sound is v. In the indicated coordinate system, the origin is located at the center of the speaker.

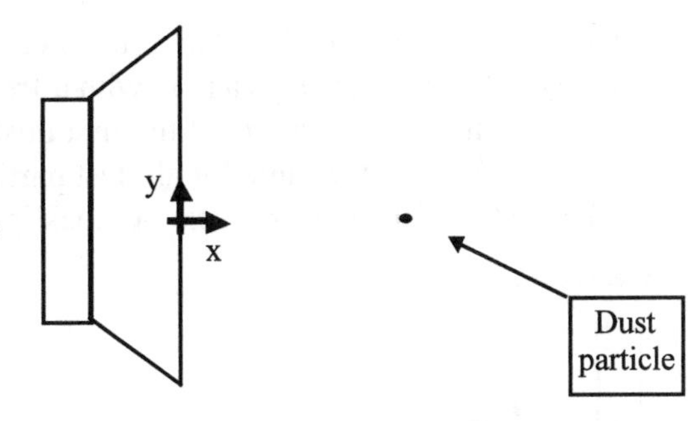

1. How long does it take for the sound wave to reach the dust particle? Explain.

2. At time t = 0 sec, the dust particle begins to move away from the loudspeaker. Write an equation that describes the displacement of the dust particle from equilibrium for all times after t = 0 sec. Explain how you arrived at your answer. Explicitly define any variables you introduce in your equation.

3. Consider an identical dust particle a distance x_0 from an identical loudspeaker. The loudspeaker is turned on and produces a sound with a frequency of $2f$. Does the dust particle begin to move earlier than in question 1? Explain.

4. Write an equation that describes the displacement of the dust particle in question 3 from equilibrium for all times after t = 0 sec. Explain how you arrived at your answer.

B. Five dust particles are placed in a row 5 cm apart beginning 50 cm from a loudspeaker (see figure). The speaker plays a note with a frequency of 1700 Hz. The speed of sound is 340 m/s. The maximum displacement of the first dust particle is $s_{max} = 3$ mm. Assume that the intensity of the sound wave is the same for all dust particles. In the indicated coordinate system, the origin is at the center of the loudspeaker. A clock is started at an arbitrary time.

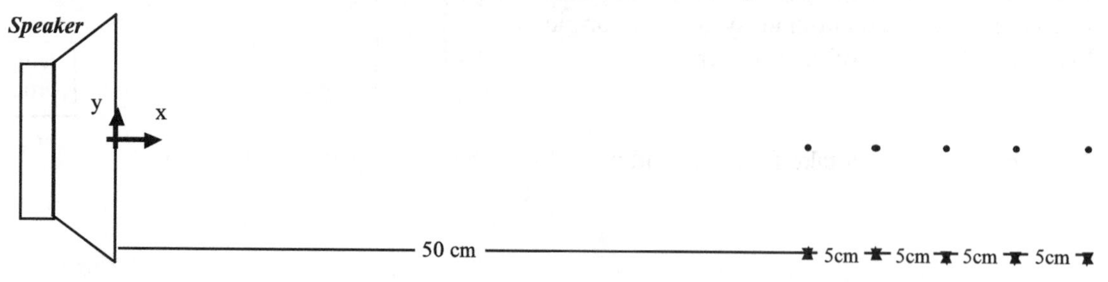

1. At time t = 0 sec, the first dust particle is at equilibrium and moving away from the loudspeaker. Find t_0, the amount of time that elapses until the second dust particle is at its equilibrium position. Explain.

2. What is the displacement from equilibrium of the first dust particle at time t_0? Explain how you arrived at your answer.

3. In the graph below, sketch a graph of s vs. x at time t_0. Define each axis clearly.

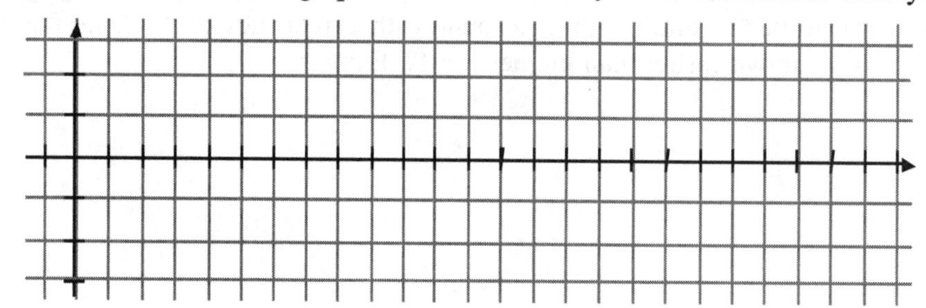

4. Find s(x,t) for $x = 65$ cm and $t = 2.941176 \times 10^{-4}$ sec. Show all work.

A. Consider the following experiment. Suppose we mix 4 g of hot water and 3 g of cold water. 12 calories of heat are transfered in the process.

1. How much heat does each gram of hot water lose? How much does the temperature of each gram of hot water change? How much does the temperature of the entire sample of hot water change? Explain how you know.

2. How much heat does each gram of cold water gain? How much does the temperature of each gram of cold water change? How much does the temperature of the entire sample of cold water change? Explain.

3. How can you account for the fact that the temperature change of the cold water differs from the temperature change of the hot water?

B. Consider an experiment in which you combine water and iron. Suppose 8 g of hot iron is placed in contact with 5 g of cold water (as shown in the figure). 10 calories of heat are transfered in the process.

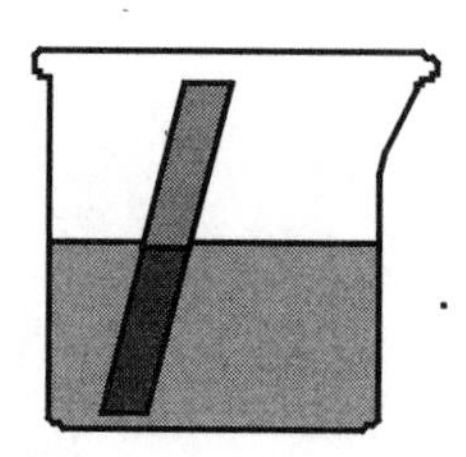

1. The specific heat of iron is about 0.12 Kcal/kg °C. How much iron has the same heat capacity as one gram of water? Explain how you arrived at your answer.

2. How much heat does each gram of hot iron lose? How much does the temperature of each gram of hot iron change? How much does the temperature of the entire sample of hot iron change? Explain your reasoning.

3. How much heat does each gram of cold water gain? How much does the temperature of each gram of cold water change? How much does the temperature of the entire sample of cold water change? Explain your reasoning.

C. When solving homework problems, you are often not given the temperatures of the hot and cold materials. Explain how you could calculate the temperature changes even though you did not know the starting temperatures?

A. Consider a particle of mass m that is traveling directly toward a wall (see figure). At the instant shown, the particle is moving at velocity v toward the right.

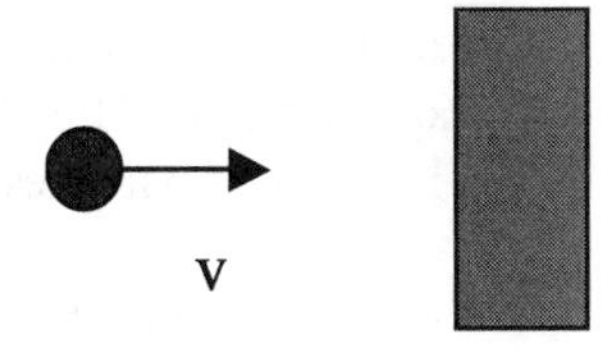

Assume that the particle strikes the wall in an elastic collision (i.e. it loses no energy in the collision). What is its change in momentum? Explain.

B. Consider a slightly modified situation. Suppose a particle of mass m strikes the wall at an angle θ and bounces off as shown in the figure at the right. Assume the particle is moving with a velocity v, and strikes the wall elastically so that it loses no energy in the collision.

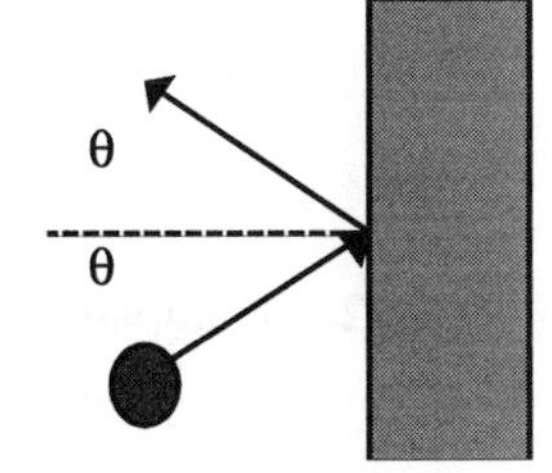

Calculate how much its momentum has changed in the collision. Explain.

C. Suppose N particles hit the wall in a small time Δt. For simplicity, assume they all move with velocity v toward the wall when they hit the wall head on ($\theta=0$) like in question A.

1. What is the total change in momentum of the gas particles that have hit the wall? Explain how you arrived at your answer.

2. What is the force that the wall has exerted on the gas molecules in question 1? Give both direction and magnitude. How do you know?

3. What is the force that the gas molecules have exerted on the wall? Give both direction and magnitude. How do you know?

D. Consider two situations where in each N particles hit a wall in a small time Δt. For simplicity, assume all particles are moving with velocity v when they hit the wall head on ($\theta=0$) like in questions 1. The two walls are of different sizes: wall A has *twice* the area of wall B. The same number of particles hits each wall.

1. Compare the change in momentum of a single particle that hits wall A and a single particle that hits wall B. Explain.

2. Compare the total change in momentum for *all N* particles that hit wall A in time Δt to the total change in momentum of *all N* particles that hit wall B in time Δt. Explain.

3. What is the force exerted on wall A compared to the force exerted on wall B? Explain.

4. Compare the pressure on wall A to the pressure on wall B. Explain how you arrived at your answer.

A. Consider the situation shown at right.

1. Draw a vector at each "x" to indicate the strength and direction of the electric field.

2. Consider the "x" located 2 grid points away from the point charge. Give *three* different possibilities for where you could place a charge in order to make the electric field at this "x" equal zero. Explain how you arrived at your answer. Be sure to indicate the magnitude and sign of the charge that you would use.

B. Consider the situation shown at right.

1. Draw vectors to indicate the electric field at each location "x." Explain how you arrived at your answer.

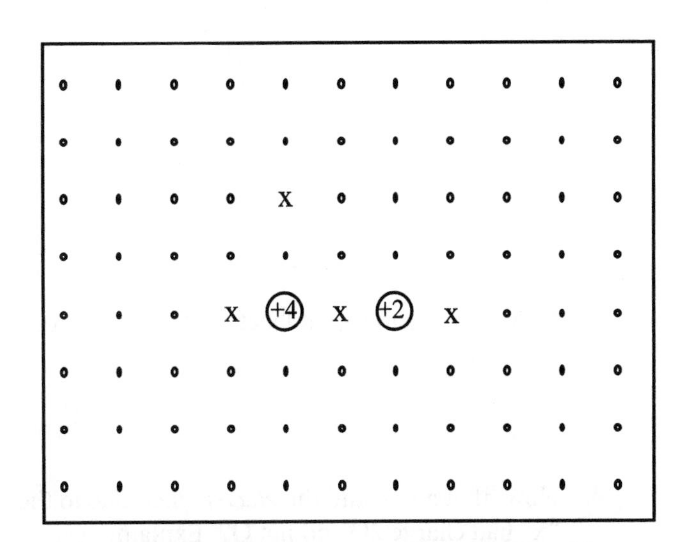

2. Consider a charge of q = +1 units in the place of the right charge. How, if at all, does this change the electric field at the point "x" located directly between the two charges? Explain.

C. Consider a set of three charges placed as shown in the diagram. The distance between each grid point is a.

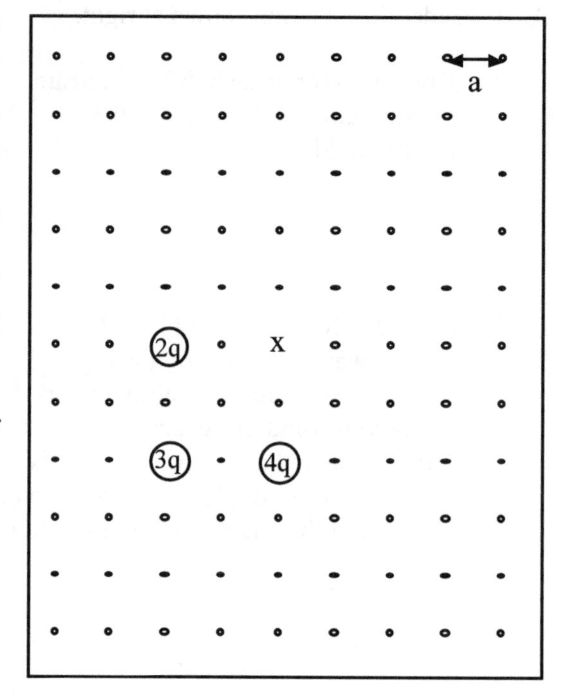

1. On the diagram, draw a separate vector to indicate the electric field due to each charge at location "x." Use a differently colored pencil for each, if possible.

   Then, sketch a vector to indicate the total electric field at location "x." Clearly label each vector. Explain how you arrived at your answer.

2. Write an equation in terms of q and a for the electric field, E_x, due to the three charges at the location "x."

3. Consider a charge Q at location "x." What force would this charge feel?

4. How, if at all, would the *electric field* due to the three point charges change if the charge at "x" had charge 2Q and not Q? Explain.

5. If a = 1 cm and q = 0.1 μC, find the electric field due to the three point charges at the location "x." Show all work.

Two charges are held fixed on a coordinate grid, as shown. The origin (0,0) is where the two axes cross. The left charge is at $\mathbf{R}_1$=(-2,0) and the right charge is at position $\mathbf{R}_2$=(2,0), where we are using the notation **R**=(x,y) to designate where to find points on the grid. Assume that charge #1 has a value $Q_1 = 1/(8.99\text{X}10^9)$ C and charge #2 has a value $Q_2 = -1/(8.99\text{X}10^9)$ C.

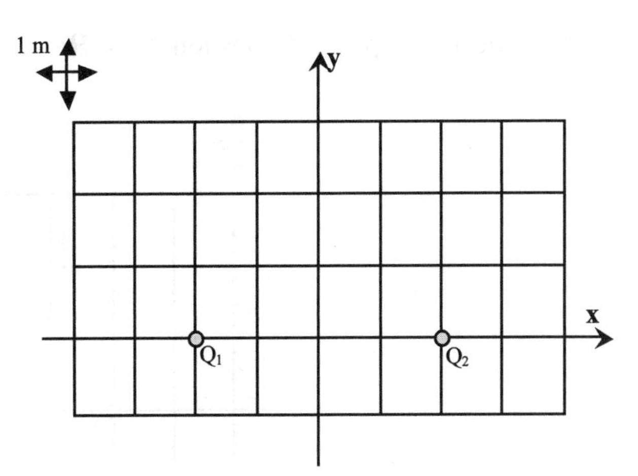

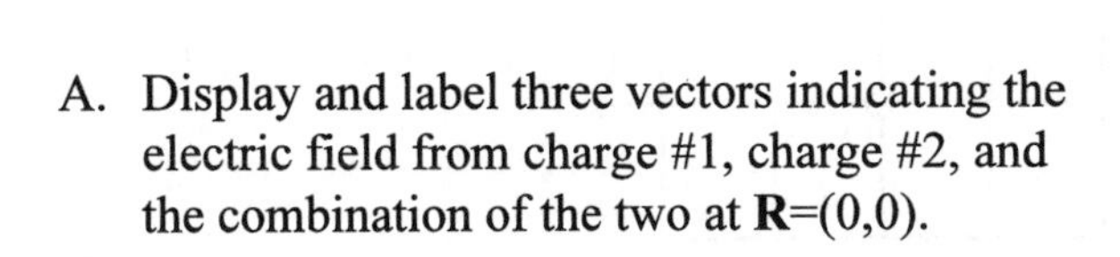

A. Display and label three vectors indicating the electric field from charge #1, charge #2, and the combination of the two at **R**=(0,0).

Calculate the total electric field at **R**=(0,0) in vector form.

2. Repeat both parts of question A for **R**=(0,2).

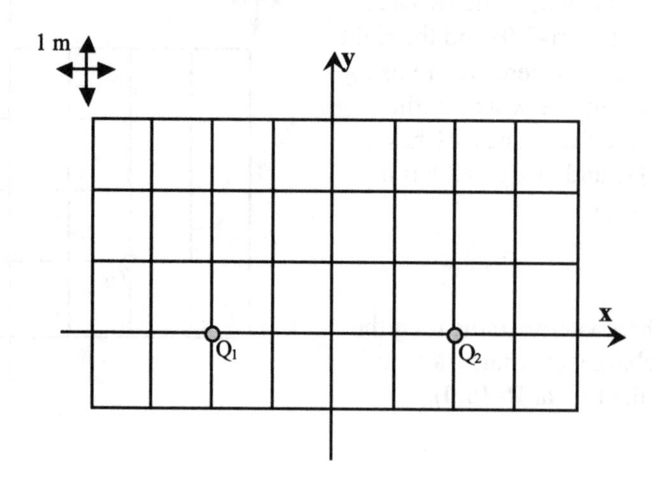

3. Repeat both parts of question A for **R**=(2,4).

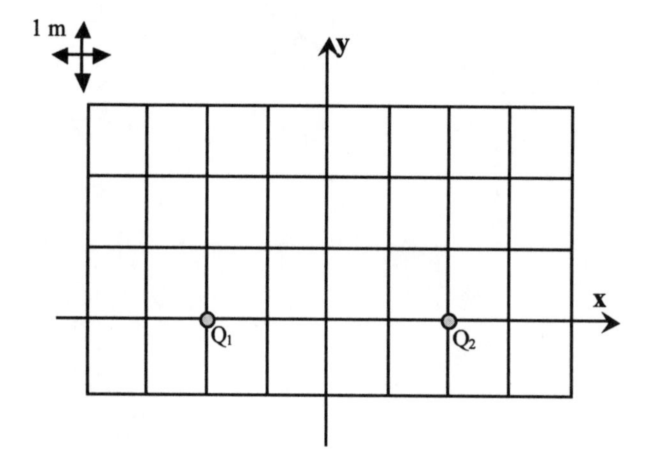

Consider the circuit shown at right. Assume that the resistance across the solenoid and the wires is zero and that the battery is ideal.

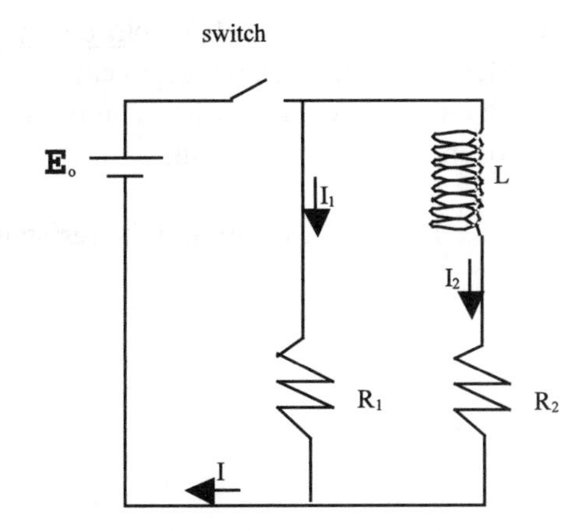

A. Just after the switch has been closed, determine the following. In each case explain how you determined your answer.

1. the current I_1 through the resistance R_1.

2. the current I_2 through the resistance R_2.

3. the total current I through the battery.

4. the potential difference across R_2.

5. the potential difference across R_1.

6. the potential difference across L.

7. dI_2/dt

(A similar problem is posed in Homework and Test Questions for Introductory Physics Teaching, Arnold Arons, Wiley, NY, 1994, p. 129.)

B. After the switch has been closed for a long time, determine the following. In each case explain how you determined your answer. The figure from page 125 has been reproduced at right.

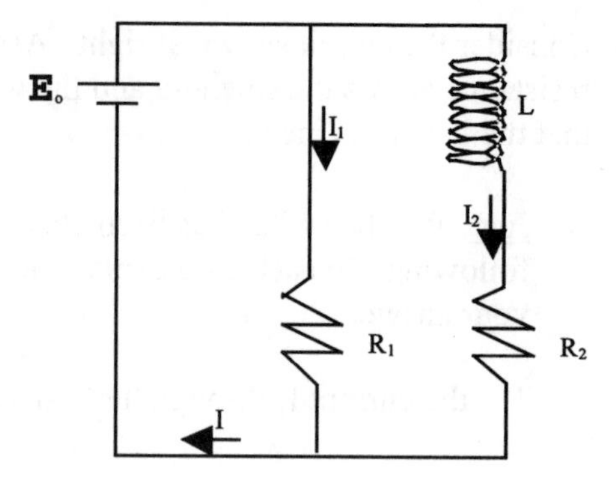

1. the current I_1 through the resistance R_1.

2. the current I_2 through the resistance R_2.

3. the total current I through the battery.

4. the potential difference across R_2.

5. the potential difference across R_1.

6. the potential difference across L.

7. dI_2/dt